THERMODYNAMICS AND HEAT ENGINES

THERMODYNAMICS AND HEAT ENGINES

Manish Saxena

ANMOL PUBLICATIONS PVT. LTD.
NEW DELHI-110 002 (INDIA)

ANMOL PUBLICATIONS PVT. LTD.
Regd. Office: 4360/4, Ansari Road, Daryaganj,
New Delhi-110002 (India)
Tel.: 23278000, 23261597, 23286875, 23255577
Fax: 91-11-23280289
Email: anmolpub@gmail.com
Visit us at: www.anmolpublications.com

Branch Office: No. 1015, Ist Main Road, BSK IIIrd Stage
IIIrd Phase, IIIrd Block, Bangalore-560 085 (India)
Tel.: 080-41723429 • Fax: 080-26723604
Email: anmolpublicationsbangalore@gmail.com

Thermodynamics and Heat Engines

First Edition, 2010

ISBN 978-81-261-4609-3

PRINTED IN INDIA

Printed at Mehra Offset Press, Delhi.

Contents

molecules at different temperatures for which these properties are not given in the literature.

METHOD OF CALCULATION

A diatomic molecule is associated with translational, rotational, vibrational and electronic motions. Corresponding to these four types of motions, there are four types of energy: translational, rotational, vibrational and electronic energy. Translational motion is due to the three dimensional movement of a molecule in space. Rotational motion is due to the rotation of the molecule as a whole about an axis passing through the centre of gravity and perpendicular to the internuclear axis. In diatomic molecules, atoms are also able to vibrate relative to each other along the internuclear axis and this is the origin of vibrational motion.

Motion of electrons in one atom is perturbed by electronic and nuclear motion in the other atom. Due to this, reshuffling of orbitals takes place and that generates molecular orbitals. This phenomenon is responsible for electronic motion. The electronic energy, <“ 1 eV to 10 eV, is very high compared to vibrational energy, <“ $10^{''2}$ eV, rotational energy, <“ $10^{''3}$eV, and translational energy, <“ $10^{''22}$ eV. However, theory shows that below 3000 K molecules are not excited electronically, and electronic motion only plays a significant role above 3000 K. Therefore electronic motion can be neglected below 3000 K.

RESULTS AND DISCUSSION

The calculated thermodynamic properties namely Gibbs energy, enthalpy, entropy and heat capacity at constant pressure, of PtH, PtC, PtN and PtO molecular gases have been estimated from spectroscopic data and are collected inTables. The spectroscopic constants which were used for the calculation of these properties are displayed in Table V (20). PtH is different from PtC, PtN and PtO since it has a $^2\Delta$ ground state. Therefore we have incorporated the ground state multiplicity in our calculation for the improvement of the results. We have estimated the thermodynamic properties from theoretical spectroscopic data and experimental spectroscopic data for the PtH molecule as shown inTable VI.

From comparison, it is clear that the Gibbs energy has a maximum deviation of 0.12%, enthalpy has 0.11%, entropy has 0.17% and heat capacity has 0.63% up to 2000 K. Ideally it is assumed that rotational and vibrational motions are independent of each other, but in practice they interact with each other. In the present calculation we include this effect by taking the vibrational-rotational partition function instead of the independent rotational and vibrational partition functions.

This gives more accurate values of thermodynamic properties than the values obtained from individual rotational and vibrational partition functions. A similar approach has been applied for the calculation of thermodynamic properties of monohalides of potassium, and the obtained results were in close agreement with reported values. Accuracy of the data also depends on the vibrational-rotational coupling. If coupling is weak, the stretching constant (α_e) is sufficient for the calculation of thermodynamic properties. If coupling is strong, the incorporation of the vibration-rotation constant ($\tilde{a}_e$) gives more accurate data.

SECOND LAW OF THERMODYNAMICS

LAWS OF HEAT POWER

The Second Law of Thermodynamics is one of three Laws of Thermodynamics. The term "thermodynamics" comes from two root words: "thermo," meaning heat, and "dynamic," meaning power.

Thus, the Laws of Thermodynamics are the Laws of "Heat Power." As far as we can tell, these Laws are absolute. All things in the observable universe are affected by and obey the Laws of Thermodynamics.

The First Law of Thermodynamics, commonly known as the Law of Conservation of Matter, states that matter/energy cannot be created nor can it be destroyed. The quantity of matter/energy remains the same. It can change from solid to liquid to gas to plasma and back again, but the total amount of matter/energy in the universe remains constant.

INCREASED ENTROPY

The Second Law of Thermodynamics is commonly known as the Law of Increased Entropy. While quantity remains the same (First Law), the quality of matter/energy deteriorates gradually over time. How so? Usable energy is inevitably used for productivity, growth and repair. In the process, usable energy is converted into unusable energy. Thus, usable energy is irretrievably lost in the form of unusable energy. "Entropy" is defined as a measure of unusable energy within a closed or isolated system (the universe for example). As usable energy decreases and unusable energy increases, "entropy" increases. Entropy is also a gauge of randomness or chaos within a closed system. As usable energy is irretrievably lost, disorganization, randomness and chaos increase.

BEGINNING OF LAW

The implications of the Second Law of Thermodynamics are considerable. The universe is constantly losing usable energy and never gaining. We logically conclude the universe is not eternal. The universe had a finite beginning — the moment at which it was at "zero entropy" (its most ordered possible state). Like a wind-up clock, the universe is winding down, as if at one point it was fully wound up and has been winding down ever since. The question is who wound up the clock?

The theological implications are obvious. Astronomer Robert Jastrow commented on these implications when he said, "Theologians generally are delighted with the proof that the universe had a beginning, but astronomers are curiously upset. It turns out that the scientist behaves the way the rest of us do when our beliefs are in conflict with the evidence."

ENTROPY

The concept of entropy is fundamental to understanding

the second law of thermodynamics. Entropy (or more specifically, increase in entropy) is defined as heat (in calories or Btu's) absorbed by a system, divided by the absolute temperature of the system at the time the heat is absorbed. Absolute temperature is the number of degrees above "absolute zero", the coldest temperature that can exist.

The total entropy in a system is represented by the symbol S. The symbol S is used to represent a given change in the entropy content of a system. If the symbol q is used to represent the amount of heat absorbed by a system, the equation for the resulting entropy increase is:

$$S = q/T \quad (1)$$

Where T is the absolute temperature. When heat is absorbed, the entropy of a system increases; when heat flows out of a system, its entropy decreases.

The "surroundings" of a system is everything outside of the system that can interact with it; surroundings can usually be defined as the space that surrounds a system. When heat is evolved by a system, that same heat is absorbed by its surroundings. When heat is absorbed by a system, that same heat must necessarily come from its surroundings. Therefore any entropy increase in a system due to heat flow must be accompanied by an entropy decrease in the surroundings, and vice versa. When heat flows spontaneously from a hotter region to a cooler region, the entropy decrease in the hotter region will always be less than the entropy increase in the cooler region, because the greater the absolute temperature, the smaller the entropy change for any particular heat flow.

As an example, consider the entropy change when a large rock at 500 degrees absolute is dropped into water at 650 degrees absolute. (We are using an absolute temperature scale based on Fahrenheit degrees; on this scale, water freezes at 492 degrees.) For each Btu of heat that flows into the rock at these temperatures the entropy increase in the rock is 1/500 = 0.0020 and the entropy decrease of the water is 1/650 = 0.0015. The difference between these values is 0.0020 - 0.0015 = 0.0005. This represents the *over all* entropy increase of the system (rock) and its surroundings (water).

Of course the rock will warm up to, and the water cool to, a temperature intermediate between their original temperatures, thus considerably complicating the calculation of total entropy change after equilibrium is achieved. Nevertheless, for every Btu of heat transferred from water to rock there will always be an increase of over-all net entropy.

As was mentioned before, a spontaneous change is an irreversible change. Therefore an increase in the overall net entropy can be used as a measure of the irreversibility of spontaneous heat flow.

Irreversible changes in a system can, and often do, take place even though there may be no interaction, and negligible heat flow, between system and surroundings. In cases like these the entropy "content" of the system is greater after the change than before. Even when heat flow does not occur between system and surroundings, spontaneous changes inside an isolated system are always accompanied by an increase in the system's entropy, and this calculated entropy increase can be used as a measure of irreversibility. The following paragraphs will explain how this entropy increase can, at least in some cases, be calculated.

It is an axiom of thermodynamics that entropy, like temperature, pressure, density, etc., is a property of a system and depends only on the existing condition of the system. Regardless of the procedures followed in achieving a given condition, the entropy content for that condition is always the same. In other words, for any given set of values for pressure, temperature, density, composition, etc., there can be only one value for the entropy content. It is essential to remember this: When a system that has undergone an irreversible change is restored to its original condition (same temperature, pressure, volume, etc.) its entropy content will likewise be the same as it was before.

In cases where an isolated system undergoes an entropy increase as the result of a spontaneous change inside the system, we can calculate that entropy increase by postulating a procedure whereby the system's entropy increase is transferred to the surroundings in a manner such that there is no further increase in *net* entropy and the system is restored to its original condition. The entropy increase of the surroundings can then be readily calculated by equation (1):

$$S = q/T,$$

where

q = heat absorbed by the surroundings

T = absolute temperature of the surroundings.

It bears repeating that when the system is restored to its original condition, its entropy content will be the same as it was before its irreversible change. Therefore the amount of entropy absorbed by the surroundings during restoration must necessarily be the same as the entropy increase accompanying the system's original irreversible change, providing that there is no further increase in *net* entropy during restoration.

This postulated restoration procedure and the postulated properties of the surroundings are for the purpose of calculation only. Since we are not dealing with the surroundings as such, they can be postulated in whatever form necessary to simplify the calculations; it is neither necessary nor desirable that the surroundings correspond to any condition that could actually exist. Therefore, we will postulate a theoretical restoration procedure that takes place with no further increase in *net* entropy, even though such a procedure can not actually be obtained experimentally.

The restoration process, if it were to take place in actuality, would have to be accompanied by at least a small amount of irreversibility, and hence an additional increase in the entropy of the surroundings beyond the entropy increase from the system's original irreversible change. This is because heat will not flow without a temperature differential, friction cannot be entirely eliminated, etc. Therefore the restoring process, if it is to take place with no further increase in overall net entropy, must be postulated to take place with no irreversibility. If such a process could be actually realized, it would be characterized by a continuous state of equilibrium (i.e. no pressure or temperature differentials) and would occur at a rate so slow as to require infinite time. Processes like these are called "reversible" processes. Remember, reversible processes are postulated to simplify the calculation of the entropy change in a system; it is *not* necessary that they be capable of being achieved experimentally.

It should not be assumed that equation (1) requires that q, the heat absorbed, must necessarily be absorbed reversibly. The concept of reversibility is merely a means to an end: the calculation of entropy change accompanying an irreversible process.

The following example will illustrate the calculation of a reversible restoring process and at the same time develop the equation which is the basis for the thermodynamical relationship between probability and the second law. We will postulate a system consisting of an "ideal" gas contained in a tank connected to a second tank that has been completely evacuated, with the valve between the two tanks closed. The temperature of the system and its surroundings is postulated to be the same.

An ideal gas is one whose molecules are infinitely small and have no attractive or repulsive forces on each other. (Under ordinary conditions hydrogen and helium closely approximate the properties of an ideal gas.) An ideal gas is chosen in order to develop the basic relationship without introducing complicating correction factors to account for the size of the molecules and the forces they exert on each other.

When the valve is opened the gas expands irreversibly from V1 (its original volume) to V2 (the volume of both tanks). There is no work of compression by or upon the surroundings. Because the gas is ideal there is no temperature change, and hence no heat flow takes place.

After expanding irreversibly from V1 to V2, the gas is restored to its original condition by reversibly compressing it back to V1. This compression requires work (force applied through a distance) which in turn generates heat in the gas, heat that is absorbed by the surroundings so that there is no increase in the gas temperature. In our mathematical model of this reversible restoring process, the surroundings are postulated to be so large that they also do not undergo any temperature increase. The temperature T remains unchanged during the entire irreversible expansion and subsequent reversible restoration process.

The work of compressing the gas during restoration is equal to the pressure of the gas times the volume change due to compression. Because the pressure increases during compression, the work of compression must be determined by the calculus integral:

compression work = PdV

where: P = pressure

V = volume

dV = the small change in volume taking place at the corresponding pressure P

The integral sign indicates the summation of all the individual values of PdV.

The equation relating temperature, pressure, and volume of an ideal gas is:

$$PV = RT \quad (2)$$

where,

P = pressure

V = volume

T = absolute temperature

R = a constant which depends only on the amount of gas present

In the case of a reversible, isothermal compression of an ideal gas we may substitute P from equation (2) into the equation for compression work. When this is done, we have:

$$\text{compression work} = RTdV/V \quad (3)$$

Although it is not necessary that our postulated reversible restoration process be capable of being carried out in a practical sense, it is nevertheless sometimes helpful to be able to visualize the process.

To this end, the reader may consider the restoring compression process being brought about by a piston fitted into the end of the second tank. On compression from V2 to V1, the piston moves down the length of the second tank, and with no mechanical friction forces all the gas contained therein back into the first tank V1.

Since the work of compression is equal to q, the heat absorbed by the surroundings, q may be substituted in equation (3) to give:

$$q = RTdV/V \quad (4)$$

From equation (1) the entropy gained by the surroundings during restoration from V2 to V1 is:

$\Delta S = q/T$ (1)

Substituting from equation (4):

$\Delta S = RdV/V$

Upon integrating (a calculus procedure for summing up the individual values of dV/V) we have:

$$DS = R\ln(V2/V1) \quad (5)$$

Where ln(V2/V1) is the natural logarithm of the ratio of expanded volume to the initial volume, and ΔS is equal to the entropy increase in the surroundings upon restoration compression from V2 to V1. As we have seen, ΔS is also equal to the entropy increase of the gas caused by its original expansion from V1 to V2. This is because V1 is the same volume both before expansion and after restoration compression, and therefore has the same entropy content. Therefore the entropy transferred to the surroundings during restoration is equal to that gained by the system in expanding from V1 to V2.

ENTROPY AND PROBABILITY

The ratio of the probability that all the gas molecules are evenly distributed between the two tanks to the probability that all the molecules, of their own accord and by random motion, would be in tank V1 is equal to $(V2/V1)^N$, where N is the number of molecules.

If V2/V1 were equal to 2.0, for example, and N were equal to 10, the probability ıatio would be 2 to the tenth power, or 1024. For N = 100, the ratio would be approximately 1.27 times ten to the 30th power. It is clear that the random motion of trillions of gas molecules heavily favours a uniform distribution.

let X1 = the probability of all the gas molecules, after the valve is opened, remaining in the first tank V1 let X2 = the probability of all the gas molecules, after the valve is opened, being uniformly distributed in V2, the volume of both tanks.

From the probability equation, we have:

$X2/X1 = (V2/V1)^N$.

Taking the natural logarithm of both sides, and then multiplying both sides by R, the gas constant:

$R \ln(X2/X1) = RN \ln(V2/V1)$

$R/N \ln(X2/X1) = R \ln(V2/V1)$

Substituting in equation (5):

$$DS = R/N \ln(X2/X1) \quad (6)$$

Equation (6) represents the fundamental relationship between probability and the second law of thermodynamics. It states that the entropy of a gaseous system increases when its molecular distribution changes from a lower probability to a higher probability (X2 greater than X1).

Based on the belief that the laws of thermodynamics are universal, this equation has been assumed to apply to all systems, not just gaseous. In other words, any entropy change is proportional to the logarithm of the ratio of probabilities. Therefore, for the general case equation (6) can be written:

$$DS = K \ln(X2/X1) \quad (7)$$

Where K is a constant depending on the particular change involved. However, individual values of K, X1, or X2 are seldom, if ever, known for non-gaseous systems.

As we have seen before, ΔS can be either positive or negative.

When ΔS is negative equation (7) can be written:

$-\Delta S = -K \ln(X2/X1)$

$= K \ln(X1/X2)$

Therefore a system can go from a more probable state (X2) to a less probable state (X1), providing S for the system is negative. In cases where the system interacts with its surroundings, ΔS can be negative *providing* the over-all entropy of the system and its interacting surroundings is positive; the over-all change can be positive if the entropy increase of the surroundings is numerically greater than the entropy decrease of the system.

In the case of the formation of the complex molecules characteristic of living organisms, creationists raise the point that when living things decay after death, the process of decay takes place with an increase in entropy. They also point out, correctly, that a spontaneous change in a system takes place with a high degree of probability. They fail to realize, however, that probability is relative, and a spontaneous change in a system can be reversed, providing the system interacts with its surroundings in such a manner that the entropy increase in the surroundings is more than enough to reverse the system's original entropy increase.

The application of energy can reverse a spontaneous, thermodynamically "irreversible" reaction. Leaves will spontaneously burn (combine with oxygen) to form water and carbon dioxide. The sun's energy, through the process of photosynthesis, will produce leaves from water vapour and carbon dioxide, and form oxygen.

If we unplug a refrigerator, heat will flow to the interior from the surroundings; the entropy increase inside the refrigerator will be greater than the entropy decrease in the surroundings, and the net entropy change is positive. If we plug it in, this spontaneous "irreversible" change is reversed. Due to the input of electrical energy to the compressor, the heat transferred to the surroundings from the condenser coils is greater than the heat extracted from the refrigerator, and the entropy increase of the surroundings is greater than the entropy decrease of the interior, in spite of the fact that the surroundings are at a higher temperature. Here again, the net entropy change is positive, as would be expected for any spontaneous process.

In a similar manner, the application of electrical energy can reverse the spontaneous reaction of hydrogen and oxygen to form water: when a current is passed through a water solution, hydrogen is liberated at one electrode, oxygen at the other.

As can easily be confirmed experimentally, agitating water raises its temperature. When water falls freely from a higher elevation to a lower elevation, its energy is changed from potential to kinetic, and finally to heat as it splashes at the end of its fall. The second law of thermodynamics states that the water will not spontaneously raise itself to its original elevation using the heat produced on splashing as the sole source of energy. To do so would require a heat engine that would convert all of the heat of splashing to mechanical energy.

The efficiency of a heat engine is thermodynamically limited by the Carnot cycle, which limits the efficiency of any heat engine to $\Delta T/T$, where T is the temperature increase due to splashing, and T is the absolute temperature. Since ΔT is only a small fraction of ΔT, there is no device that could be constructed which would allow all of the water to spontaneously jump back to its former elevation.

We can, at least in theory, calculate the entropy increase of the water resulting from its irreversible change in falling. In a manner analogous to that used in the previous example, the entropy increase would be equal to the heat generated by splashing agitation, divided by the absolute temperature. If some of the energy of the falling water is extracted by a water wheel, there will be less heat of splashing and hence less entropy increase. A properly designed turbine could extract most of the water's kinetic energy. This is *not* the same thing as trying to utilize the heat of splashing as an energy source for a heat engine to raise the water. In other words, using the energy before it becomes heat is much more efficient than trying to use it after it becomes heat.

If a water wheel is connected by shafts, belts, pulleys, etc. to a pump, the pump can raise water from the downstream side of the water wheel to an elevation even higher than that of the upstream reservoir. *Some* of the water would spontaneously raise itself to an elevation even higher than original, but the rest of it would end up below the water wheel on the downstream side.

While it is not possible for *all* of the water to raise itself to an elevation higher than its initial elevation, it is possible for *some* of the water to spontaneously raise itself to an elevation higher than initial.

As with any other irreversible change, there will be an increase in over-all entropy. This means that the entropy increase of the water going over the wheel is greater than the entropy decrease of water pumped up to the higher elevation.

This will be confirmed mathematically in the following paragraphs. will stand for the Greek letter gamma, representing unit weight in pounds per cubic foot. An increase in the value of a parametre will be represented by Δ.

Let: γ = unit weight of water, lbs./cubic foot

h = height of reservoir above downstream side, feet

γh = potential energy of water in reservoir

h = additional height above reservoir

$h + \Delta h$ = height to which water is pumped, feet above downstream side

w = work of pumping water to higher elevation

q = heat loss due to pump friction and downstream agitation

f = fraction of water pumped to elevation $h + \Delta h$

From the flow equation, energy in = energy out:

$$gh = q + w$$

Where: γh = the total energy expended by the water in falling from height h.

q = the energy wasted as heat of splashing

w = work expended in pumping the water to a higher elevation

$q = T\Delta S$ (from equation 1)

$$w = fg(h + Dh)$$

The total energy available, h, is divided into pump work, $f\gamma(h + \Delta h)$, and energy lost, TS:

$$gh = TDS + fg(h + Dh)$$

Rearranging:

$$TDS = gh - fg(h + Dh) \quad (8)$$

When no pump work is done, then:

$$TDS' = gh \quad (9)$$

Combining equations (8) and (9), we get:

$$DS' - DS = fg(h + Dh)/T \quad (10)$$

In the case where the water falls freely without turning the water wheel or operating the pump:

let q'= heat of splashing

$q' = T\Delta S'$

where $\Delta S'$ is the entropy increase on free fall

Equation (10) shows that $\Delta S'$ is larger than ΔS, and that the entropy increase due to pump friction and downstream agitation is "backed up" by the even larger entropy increase that takes place when water falls freely. Equation (10) also shows that the lower the value of ΔS, the more efficient the pump, and the greater the value of f, the fraction of water pumped.

Creationists assume that a change characterized by a decrease in entropy can not occur under any circumstances. In fact, spontaneous entropy decreases can, and do, occur all the time, providing sufficient energy is available. The fact that the water wheel and pump are man-built contraptions has no bearing on the case: thermodynamics does not concern itself with the detailed description of a system; it deals only with the relationship between initial and final states of a given system (in this case, the water wheel and pump).

A favourite argument of creationists is that the probability of evolution occurring is about the same as the probability that a tornado blowing through a junkyard could form an airplane. They base this argument on their belief that changes in living things have a very low probability and could not occur without "intelligent design" which overcomes the laws of thermodynamics. This represents a fundamental contradiction in which (they say) evolution is inconsistent with thermodynamics because thermodynamics doesn't permit order to spontaneously arise from disorder, *but* creationism (in the guise of intelligent design) doesn't have to be consistent with the laws of thermodynamics.

A simpler analogy to the airplane/junkyard scenario would be the stacking of three blocks neatly on top of each other. To do this, intelligent design is required, but stacking does not violate the laws of thermodynamics. The same relations hold for this activity as for any other activity involving thermodynamical energy changes. It is true that the blocks will not stack themselves, but as far as *thermodynamics* is concerned, all that is required is the energy to pick them up and place them one on top of the other. Thermodynamics merely correlates the energy relationships in going from state A to state B. If the energy relationships permit, the change may occur. If they don't permit it, the change can not occur. A ball will not spontaneously leap up from the floor, but if it is dropped, it will spontaneously bounce up from the floor. Whether the ball is lifted by intelligent design or just happens to fall makes no difference.

On the other hand, thermodynamics does not rule out the possibility of intelligent design; it is just simply not a factor with respect to the calculation of thermodynamic probability.

Considering the earth as a system, any change that is accompanied by an entropy decrease (and hence going back from higher probability to lower probability) is possible as long as sufficient energy is available. The ultimate source of most of that energy, is of course, the sun.

The numerical calculation of entropy changes accompanying physical and chemical changes are very well understood and are the basis of the mathematical determination of free energy, emf characteristics of voltaic cells, equilibrium constants, refrigeration cycles, steam turbine operating parametres, and a host of other parametres. The creationist position would necessarily discard the entire mathematical framework of thermodynamics and would provide no basis for the engineering design of turbines, refrigeration units, industrial pumps, etc. It would do away with the well-developed mathematical relationships of physical chemistry, including the effect of temperature and pressure on equilibrium constants and phase changes.

Sometimes people say that life violates the second law of thermodynamics. This is not the case; we know of nothing in the universe that violates that law. So why do people say that life violates the second law of thermodynamics? What is the second law of thermodynamics?

The second law is a straightforward law of physics with the consequence that, in a closed system, you can't finish any real physical process with as much useful energy as you had to start with — some is always wasted. This means that a perpetual motion machine is impossible. The second law was formulated after nineteenth century engineers noticed that heat cannot pass from a colder body to a warmer body by itself.

THERMODYNAMIC ENTROPY

The first opportunity for confusion arises when we introduce the term *entropy* into the mix. Clausius invented the term in 1865. He had noticed that a certain ratio was constant in reversible, or ideal, heat cycles. The ratio was heat exchanged to absolute temperature. Clausius decided that the conserved ratio must correspond to a real, physical quantity, and he named it "entropy".

Surely not every conserved ratio corresponds to a real, physical quantity. Historical accident has introduced this term to science. On another planet there could be physics without the concept of entropy. It completely lacks intuitive clarity. Even the great physicist James Clerk Maxwell had it backward for a while . Nevertheless, the term has stuck.

The *American Heritage Dictionary* gives as the first definition of entropy, "For a closed system, the quantitative measure of the amount of thermal energy not available to do work." So it's a negative kind of quantity, the opposite of available energy.

Today, it is customary to use the term entropy to state the second law: Entropy in a closed system can never decrease. As long as entropy is defined as unavailable energy, the paraphrasing just given of the second law is equivalent to the earlier ones above. In a closed system, available energy can never increase, so (because energy is conserved) its complement, entropy, can never decrease.

A familiar demonstration of the second law is the flow of heat from hot things to cold, and never vice-versa. When a hot stone is dropped into a bucket of cool water, the stone cools and the water warms until each is the same temperature as the other. During this process, the entropy of the system increases. If you know the initial temperatures of the stone and the water, and the final temperature of the water, you can quantify the entropy increase in calories or joules per degree.

You may have noticed the words "closed system" a couple of times above. Consider simply a black bucket of water initially at the same temperature as the air around it. If the bucket is placed in bright sunlight, it will absorb heat from the sun, as black things do. Now the water becomes warmer than the air around it, and the available energy has increased. Has entropy*decreased*? Has energy that was previously unavailable become available, in a closed system? No, this example is only an apparent violation of the second law. Because sunlight was admitted, the local system was not closed; the energy of sunlight was supplied from outside the local system. If we consider the larger system, including the sun, available energy has decreased and entropy has increased as required.

LOGICAL ENTROPY

Entropy is also used to mean disorganization or disorder. J. Willard Gibbs, the nineteenth century American theoretical physicist, called it "mixedupness." The *American Heritage Dictionary* gives as the second definition of entropy, "a measure of disorder or randomness in a closed system." Again, it's a negative concept, this time the opposite of organization or order. The term came to have this second meaning thanks to the great Austrian physicist Ludwig Boltzmann.

In Boltzmann's day, one complaint about the second law of thermodynamics was that it seemed to impose upon nature a preferred direction in time. Under the second law, things can only go one way. This apparently conflicts with the laws of physics at the molecular level, where there is no preferred direction in time — an elastic collision between molecules would look the same going forward or backward. In the 1880s and 1890s, Boltzmann used molecules of gas as a model, along with the laws of probability, to show that there was no real conflict. The model showed that, no matter how it was introduced, heat would soon become evenly diffused throughout the gas, as the second law required.

The model could also be used to show that two different kinds of gasses would become thoroughly mixed. The reasoning he used for mixing is very similar to that for the diffusion of heat, but there is an important difference. In the diffusion of heat, the entropy increase can be measured with the ratio of physical units, joules per degree. In the mixing of two kinds of gasses already at the same temperature, if no energy is dissipated, the ratio of joules per degree — thermodynamic entropy — is irrelevant. The non-dissipative mixing process is related to the diffusion of heat only by analogy . Nevertheless, Boltzmann used a factor, *k*, now called Boltzmann's constant, to attach physical units to the latter situation. Now the word entropy has come to be applied to the simple mixing process, too. (Of course, Boltzmann's constant has a legitimate use — it relates the average kinetic energy of a molecule to its temperature.)

Entropy in this latter sense has come to be used in the growing fields of information science, computer science, communications theory, etc. The story is often told that in the late 1940s, John von Neumann, a pioneer of the computer age, advised communication-theorist Claude E. Shannon to start using the term "entropy" when discussing information because "no one knows what entropy really is, so in a debate you will always have the advantage" (6).

Richard Feynman knew there is a difference between the two meanings of entropy. He discussed thermodynamic entropy is called "Entropy" using physical units, joules per degree, and over a dozen equations.

So we now have to talk about what we mean by disorder and what we mean by order. Suppose we divide the space into little volume elements. If we have black and white molecules, how many ways could we distribute them among the volume elements so that white is on one side and black is on the other? On the other hand, how many ways could we distribute them with no restriction on which goes where? Clearly, there are many more ways to arrange them in the latter case. We measure "disorder" by the number of ways that the insides can be arranged, so that from the outside it looks the same. *The logarithm of that number of ways is the entropy.* The number of ways in the separated case is less, so the entropy is less, or the "disorder" is less.

This is Boltzmann's model again. Notice that Feynman does not use Boltzmann's constant. He assigns no physical units to this kind of entropy, just a number (a logarithm.)

Notice another thing. The "number of ways" can only be established by first artificially dividing up the space into little volume elements. This is not a small point. In every real physical situation, counting the number of possible arrangements requires an arbitrary parceling.

There is, however, nothing to tell us how fine the [parceling] should be. Entropies calculated in this way depend on the size-scale decided upon, in direct contradiction with thermodynamics in which entropy changes are fully objective.

This sort of entropy is clearly different. Physical units do not pertain to it, and (except in the case of digital information) an arbitrary convention must be imposed before it can be quantified. To distinguish this kind of entropy from thermodynamic entropy, let's call it *logical entropy*.

In spite of the important distinction between the two meanings of entropy, the rule as stated above for thermodynamic entropy seems to apply nonetheless to the logical kind: entropy in a closed system can never decrease. And really, there would be nothing mysterious about this law either. It's similar to saying *things never organize themselves*. (The original meaning of organize is "to furnish with organs.") Only this rule has little to do with thermodynamics.

It is true that crystals and other regular configurations can be formed by unguided processes. And we are accustomed to saying that these configurations are "organized." But crystals have not been spontaneously "furnished with organs." The correct term for such regular configurations is "ordered." The recipe for a crystal is already present in the solution it grows from the crystal lattice is prescribed by the structure of the molecules that compose it. The formation of crystals is the straightforward result of chemical and physical laws that do not evolve and that are, compared to genetic programs, very simple.

The rule that things never organize themselves is also upheld in our everyday experience. Without someone to fix it, a broken glass never mends. Without maintenance, a house deteriorates. Without management, a business fails. Without new software, a computer never acquires new capabilities.

Charles Darwin understood this universal principle. It's common sense. That's why he once made a note to himself pertaining to evolution, "Never use the words higher or lower".

Even today, if you assert that a human is more highly evolved than a flatworm or an *amoeba,* there are darwinists who'll want to fight about it. They take the position, apparently, that evolution has not necessarily shown a trend toward more highly organized forms of life, just different forms.

LIFE IS ORGANIZATION

Seen in retrospect, evolution as a whole doubtless had a general direction, from simple to complex, from dependence on to relative independence of the environment, to greater and greater autonomy of individuals, greater and greater development of sense organs and nervous systems conveying and processing information about the state of the organism's surroundings, and finally greater and greater consciousness. You can call this direction progress or by some other name.

Life is organization. From prokaryotic cells, eukaryotic cells, tissues, and organs, to plants and animals, families, communities, ecosystems, and living planets, life is organization, at every scale. The evolution of life is the increase of biological organization, if it is anything. Clearly, if life originates and makes evolutionary progress without organizing input from outside, then something has organized itself. Logical entropy in a closed system has decreased. This is the violation that people are getting at, when they say that life violates the second law of thermodynamics. This violation, the *decrease* of logical entropy in a closed system, must happen continually in the darwinian account of evolutionary progress.

Most darwinists just ignore this staggering problem. When confronted with it, they seek refuge in the confusion between the two kinds of entropy. Entropy [logical] has not decreased, they say, because the system is not closed. Energy such as sunlight is constantly supplied to the system. If you consider the larger system that includes the sun, entropy has increased, as required.

HEAT ENGINE

Heat engine is defined as *a device that converts heat energy into mechanical energy* or more exactly *a system which operates continuously and only heat and work may pass across its boundaries.*

The operation of a heat engine can best be represented by a thermodynamic cycle. Some examples are: Otto, Diesel, Brayton, Stirling and Rankine cycles.

Forward Heat Engine

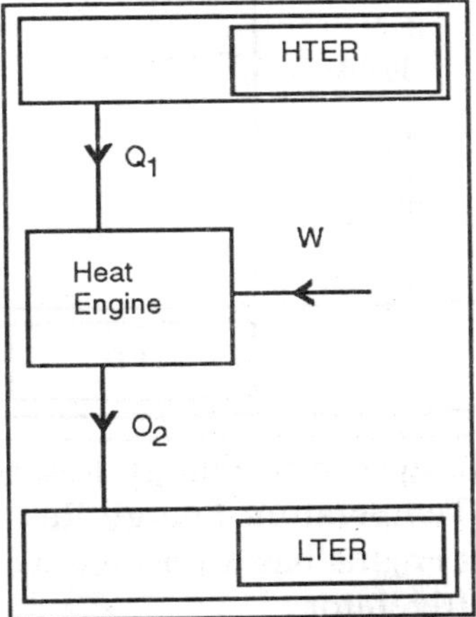

LTER= Low Temperature Energy Reservoir
HTER= High Temperature Energy Reservoir
A forward heat engine has a positive work output such

as Rankine or Brayton cycle. Applying the first law of thermodynamics to the cycle gives:

$$Q1 - Q2 - W = 0$$

The second law of thermodynamics states that the thermal efficiency of the cycle,η, has an upper limit (the thermal efficiency of the Carnot cycle), i.e.

It can be shown that:

$$Q1 > W$$

which means that it is impossible to convert the whole heat input to work and

$$Q2 > 0$$

which means that a minimum of heat supply to the cold reservoir is necessary.

Reverse Heat Engine

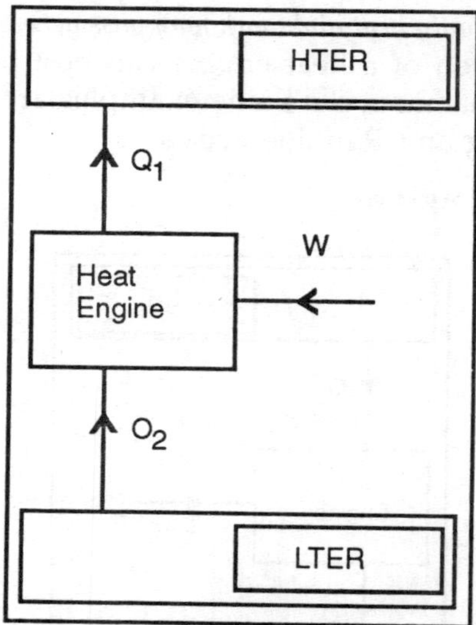

LTER= Low Temperature Energy Reservoir
HTER= High Temperature Energy Reservoir

A reverse heat engine has a positive work input such as heat pump and refrigerator.

Applying the first law of thermodynamics to the cycle gives:

$$- Q1 + Q2 + W = 0$$

In case of a reverse heat engine the second law of

thermodynamics is as follows: It is impossible to transfer heat from a cooler body to a hotter body without any work input i.e.

$$W > 0$$

REVERSIBILITY

In a reversible process the state of working fluid and system's surroundings can be restored to the original ones. This requires that the working fluid goes through a continuous series of equilibrium states.

A reversible process should satisfy the following criteria:

- No internal or mechanical friction is allowed.
- The temperature and pressure difference between the working fluid and its surroundings should be infinitely small.

There are no truly reversible processes in practice. The real processes are called irreversible. However, there are some processes that can be assumed internally reversible with good approximation, such as processes in cylinders with reciprocating piston. The working fluid is always in an equilibrium state in internally reversible process. But the surroundings undergo a state change that can never be restored.

Some processes may not be assumed internally reversible, such as processes in turbo machinery. The irreversibility of these processes are due to the high degree of turbulence of the working fluid. A reversible process between two states may be shown by a continuous curve on any diagram of properties. Different points on the curve represent the intermediate states.

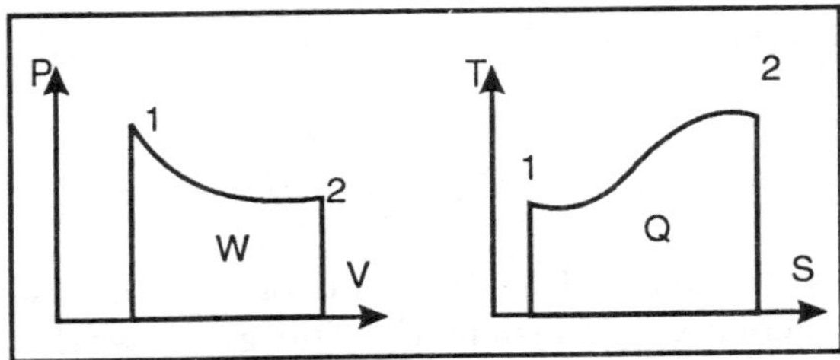

The work input to a system during a reversible process is:

W= Marked area on the P-V diagram.

and the heat supplied to a system during a reversible process is:

Q= Marked area on the T-s diagram.

Intermediate states for an irreversible process is indeterminate, therefore these processes are often shown by a dotted line joining the initial and final states.

CARNOT CYCLE

By using the second law of thermodynamics it is possible to show that no heat engine can be more efficient than a reversible heat engine working between two fixed temperature limits. This heat engine is known as Carnot cycle and consists of the following processes:

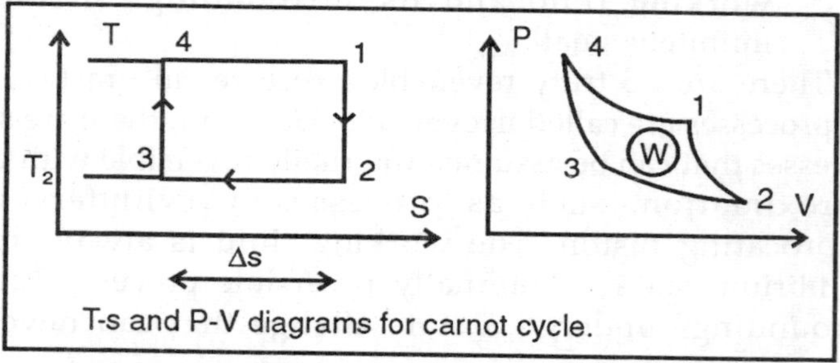

T-s and P-V diagrams for carnot cycle.

- 1 to 2: Isentropic expansion
- 2 to 3: Isothermal heat rejection
- 3 to 4: Isentropic compression
- 4 to 1: Isothermal heat supply

The supplied heat to the cycle per unit mass flow is:

Q1 = T1D s

The rejected heat from the cycle per unit mass flow is:

Q2 = T2D s

By applying the first law of thermodynamics to the cycle, we obtain:

Q1 - Q2 - W = 0

And the thermal efficiency of the cycle will be:

= hW/Q1 = 1 - T2/T1

Due to mechanical friction and other irreversiblities no cycle can achieve this efficiency. The gross work output of cycle, i.e. the work done by the system is:

$$Wg = W41 + W12$$

and work ratio is defined as the ratio of the net work, W, to the gross work output, Wg, i.e.

$$W/Wg$$

The Carnot cycle has a low work ratio. Although this cycle is the most efficient system for power generation theoretically, it can not be used in practice. There are several reasons such as low work ratio, economical aspects and practical difficulties

CLOSED AND OPEN SYSTEM

THERMODYNAMIC SYSTEM

In thermodynamics, a thermodynamic system, originally called a working substance, is defined as that part of the universe that is under consideration.

A hypothetical boundary separates the system from the rest of the universe, which is referred to as the environment, surroundings, or reservoir. A useful classification of thermodynamic systems is based on the nature of the boundary and the quantities flowing through it, such as matter, energy, work, heat, and entropy. A system can be anything, for example a piston, a solution in a test tube, a living organism, an electrical circuit, a planet, etc.

Thermodynamics is conducted under a *system-centred view* of the universe. All quantities, such as pressure or mechanical work, in an equation refer to the system unless labeled otherwise. Thermodynamics is basically concerned with the flow and balance of energy and matter in a thermodynamic system. Three types of thermodynamic systems are distinguished depending on the kinds of interaction and energy exchange taking place between the system and its surrounding environment:

- Isolated systems are completely isolated in every way from their environment. They do not exchange heat, work or matter with their environment. An example of an isolated system would be an insulated rigid container, such as an insulated gas cylinder.

- Closed systems are able to exchange energy (heat and work) but not matter with their environment. A greenhouse is an example of a closed system exchanging heat but not work with its environment. Whether a system exchanges heat, work or both is usually thought of as a property of its boundary.
- Open systems: exchanging energy (heat and work) and matter with their environment. A boundary allowing matter exchange is called *permeable*. The ocean would be an example of an open system.

In reality, a system can never be absolutely isolated from its environment, because there is always at least some slight coupling, even if only via minimal gravitational attraction. In analyzing a system in steady-state, the energy into the system is equal to the energy leaving the system.

As an example, consider the system of hot liquid water and solid table salt in a sealed, insulated test tube held in a vacuum (the surroundings). The test tube constantly loses heat (in the form of black-body radiation), but the heat loss progresses very slowly. If there is another process going on in the test tube, for example the dissolution of the salt crystals, it will probably occur so quickly that any heat lost to the test tube during that time can be neglected. (Thermodynamics does not measure time, but it does sometimes accept limitations on the timeframe of a process.)

HISTORY

The first to develop the concept of a "thermodynamic system" was the French physicist Sadi Carnot whose 1824 *Reflections on the Motive Power of Fire* studied what he called the "working substance" (system), i.e. typically a body of water vapour, in steam engines, in regards to the system's ability to do work when heat is applied to it. The working substance could be put in contact with either a heat reservoir (a boiler), a cold reservoir (a stream of cold water), or a piston (to which the working body could do work by pushing on it). In 1850, the German physicist Rudolf Clausius generalized this picture to include the concept of the surroundings, and began referring to the system as a "working body." In his 1850 manuscript *On the Motive Power of Fire,* Clausius wrote:

"With every change of volume (to the working body) a certain amount work must be done by the gas or upon it, since by its expansion it overcomes an external pressure, and since its compression can be brought about only by an exertion of external pressure.

To this excess of work done by the gas or upon it there must correspond, by our principle, a proportional excess of heat consumed or produced, and the gas cannot give up to the "surrounding medium" the same amount of heat as it receives."

The article Carnot heat engine shows the original piston-and-cylinder diagram used by Carnot in discussing his ideal engine; below, we see the Carnot engine as is typically modeled in current use: Carnot engine diagram (modern) - where heat flows from a high temperature T_H furnace through the fluid of the "working body" (working substance) and into the cold sink T_C, thus forcing the working substance to do mechanical work W on the surroundings, via cycles of contractions and expansions.

In the diagram shown, the "working body" (system), a term introduced by Clausius in 1850, can be any fluid or vapour body through which heat Q can be introduced or transmitted through to produce work. In 1824, Sadi Carnot, in his famous paper *Reflections on the Motive Power of Fire,* had postulated that the fluid body could be any substance capable of expansion, such as vapour of water, vapour of alcohol, vapour of mercury, a permanent gas, or air, etc. Although, in these early years, engines came in a number of configurations, typically Q_H was supplied by a boiler, wherein water was boiled over a furnace; Q_C was typically a stream of cold flowing water in the form of a condenser located on a separate part of the engine. The output work W here is the movement of the piston as it is used to turn a crank-arm, which was then typically used to turn a pulley so to lift water out of flooded salt mines. Carnot defined work as "weight lifted through a height."

SYSTEMS IN EQUILIBRIUM

In isolated systems it is consistently observed that as time goes on internal rearrangements diminish and stable conditions are approached. Pressures and temperatures tend to equalize, and matter arranges itself into one or a few relatively homogeneous phases. A system in which all processes of change have gone practically to completion is considered to be in a state of thermodynamic equilibrium. The thermodynamic properties of a system in equilibrium are unchanging in time. Equilibrium system states are much easier to describe in a deterministic manner than non-equilibrium states.

In thermodynamic processes, large departures from equilibrium during intermediate steps are associated with increases in entropy and increases in the production of heat rather than useful work. It can be shown that for a process to be reversible, each step in the process must be reversible. For a step in a process to be reversible, the system must be in equilibrium throughout the step. That ideal cannot be accomplished in practice because no step can be taken without perturbing the system from equilibrium, but the ideal can be approached by making changes slowly.

OPEN SYSTEMS

In open systems, matter may flow in and out of the system boundaries. The first law of thermodynamics for open systems states: the increase in the internal energy of a system is equal to the amount of energy added to the system by matter flowing in and by heating, minus the amount lost by matter flowing out and in the form of work done by the system. The first law for open systems is given by:

During steady, continuous operation, an energy balance applied to an open system equates shaft work performed by the system to heat added plus net enthalpy added. where U_{in} is the average internal energy entering the system and U_{out} is the average internal energy leaving the system. The region of space enclosed by open system boundaries is usually called a control volume, and it may or may not correspond to physical walls. If we choose the shape of the control volume such that all flow in or out occurs perpendicular to its surface, then the flow of matter into the system performs work as if it were a piston of fluid pushing mass into the system, and the system performs work on the flow of matter out as if it were driving a piston of fluid. There are then two types of work performed: *flow work* described above which is performed on the fluid (this is also often called *PV work*) and *shaft work* which may be performed on some mechanical device. These two types of work are expressed in the equation:

Substitution into the equation above for the control volume *cv* yields: The definition of enthalpy, *H*, permits us to use this thermodynamic potential to account for both internal energy and PV work in fluids for open systems:

THERMODYNAMIC POTENTIALS

Most real thermodynamic systems are open systems that exchange heat and work with their environment, rather than the closed systems described thus far. For example, living systems are clearly able to achieve a local reduction in their entropy as they grow and develop; they create structures of greater internal energy (i.e., they lower entropy) out of the nutrients they absorb. This does not represent a violation of the second law of thermodynamics, because a living organism does not constitute a closed system.

In order to simplify the application of the laws of thermodynamics to open systems, parametres with the dimensions of energy, known as thermodynamic potentials, are introduced to describe the system. The resulting formulas are expressed in terms of the Helmholtz free energy *F* and the Gibbs free energy *G*, named after the 19th-century German physiologist and physicist Hermann von Helmholtz and the contemporaneous American physicist Josiah Willard Gibbs. The key conceptual step is to separate a system from its heat reservoir. A system is thought of as being held at a constant temperature *T* by a heat reservoir (i.e., the environment), but the heat reservoir is no longer considered to be part of the system.

FLOW AND NON-FLOW PROCESSES

EFFECTS OF FLOW

Anionic surfactants, good foaming agents, can be used in CO_2 foam flooding to improve high-pressure, high-density CO_2 reservoir sweep efficiency. In this study, kinetics and equilibrium adsorption were investigated by examining adsorption behaviour in a system of solid phase sandstone or limestone and of an aqueous phase of surfactant in 2% brine. Effects on surfactant adsorption density for different solid to liquid ratios as well as surfactant concentration, rock type and state, and flow conditions are presented. Three systems were used: batch tests on crushed rock, circulation tests through core samples, and non-flow, diffusion brine-saturated core tests.

The density of an anionic surfactant adsorption on rock is best described as a function of surfactant available in the system, rather than by surfactant concentration used by previous investigators. Experiments with solid rock were carried out to determine surfactant capacity of the porous media through flow and non-flow rock samples and crushed rock samples. Adsorption was similar for crushed sandstone and flow tests, while significantly higher in cubed, non-flow rock systems. For limestone the crushed rock and non-flow systems were similar while the flow system was significantly lower. The time to reach equilibrium required less than one hour (generally minutes) for the crushed rock, hours to days for the flow-through tests, and weeks to over a month for the non-flow rock systems. The rate of adsorption dependent on availability (delivery) that is generally much slower than the adsorption kinetics.

INTRODUCTION

Surfactants are widely used in a large number of applications because of their remarkable ability to influence the properties of surfaces and interfaces. The applications of surfactants in petroleum industry are diverse and significant. Some of these areas include: in situ, wellbores, surface facilities, and environmental, health, and safety applications.[1] In each case appropriate knowledge and practices determine both the economic and technical successes of the industrial process concerned.

In-situ CO_2 flooding processes frequently experience poor sweep efficiency despite the favourable characteristics of CO_2 in recovery of oil. To mitigate this problem, significant research has focused on the use of foam in CO_2 flooding processes. It has been established that foam can improve sweep efficiency in CO_2 flooding processes by reducing the mobility of CO_2 and by diverting CO_2 flow to previously bypassed zones.[2-8]

Foam application involves injecting a surfactant along with water and gas into the reservoir. Normally, the surfactant is dissolved in the aqueous phase though surfactant dissolved in CO_2 are also being considered.[9] To aid in the selection of a suitable surfactant for the reservoir, laboratory data is collected to characterize the surfactant performance.

The economics of foam flooding depend significantly on the quantity of surfactant required to generate and propagate foam. Surfactant loss through partitioning into the crude oil phase and through adsorption onto the rock surfaces often consumes more than 90% of the surfactant in the system.[5] Surfactant loss through partitioning into the crude oil can be responsible for surfactant losses of as much as 30%. However, for the very hydrophilic surfactants chosen for many foam flooding applications, the partitioning onto crude oil is near zero.[1]

More serious losses are demonstrated from the results of a number of studies of the adsorption properties of surfactants suitable for foam flooding. These have shown that effective foam forming surfactant may exhibit adsorption levels from near zero to up to quite high levels on the order.

For example one test determined adsorption rates of 2.5 mg/g (mg of adsorbed surfactant for each gram of rock in the system). A mile-square, 10 feet thick reservoir saturated with surfactant would require 55,000 tons (50,000 tones) of surfactant at 2.5 mg of surfactant per g of rock. Adsorption does not depend on the nature of the surfactant alone, but also (though not inclusively) on temperature, brine salinity and hardness, rock type, wettability, and the presence of the residual oil phase. These factors will lead to vastly different distances of foam propagation into a reservoir, so selection of foam-forming surfactant formulation with acceptable adsorption levels at reservoir conditions is critical.

Soil respiration estimates obtained from non-flow-through steady-state chambers (also called static, absorption, or alkali trap chambers) are considered by many investigators to be unreliable. We studied the accuracy, functioning, and design requirements of this chamber type using a gas diffusion model validated for this purpose by demonstrating that it matched the empirical relation between alkali-measured flux and headspace CO_2 concentration.

Simulated measurement error depended on:

(i) Magnitude of the soil respiration rate, which spawned positive or negative error depending on the algebraic sign of the change in headspace CO_2,

(ii) Absorption efficiency of the alkali trap, which was determined by headspace air mixing rates, the thickness of atmospheric interfacial layers, and especially the ratio of exposed alkali surface area to emitting soil surface area,

(iii) The effective diffusivity and storage coefficient of CO_2 in underlying soil, which depended on the soil's air-filled porosity (AFP) and pH, respectively,

(iv) The rate of CO_2 leakage between the chamber system and its surroundings.

The results also indicated that although no single chamber design is universally applicable, striving for the ideal design in every situation is not required; for example, measurement error associated with the design used in our simulations was usually only 5% despite that headspace concentration rose more than 70% within 2 h. Larger errors occurred for chamber designs less well matched to the soil respiration rate they were intended to measure, but if such serious design deficiencies are avoided, the method offers a simple inexpensive means for obtaining multiple reliable time-integrated estimates of soil respiration, even at remote locations.

Abbreviations

AFP, air-filled porosity (m^3 m^{-3} soil):

- *D*, binary molecular diffusion coefficient of CO_2 in air
- F_a, estimate of CO_2 flux density at the soil-atmosphere boundary obtained from the amount of CO_2 absorbed by the alkali trap in a non-flow-through steady-state chamber
- F_c, CO_2 flux density at the soil-atmosphere boundary
- F_o, depth-integrated rate of subsurface CO_2 production expressed in units of flux density
- FT, flow-through
- NFT, non-flow-through
- NSS, non-steady-state
- SS, steady-state

NON-FLOW-THROUGH STEADY

STATE (NFT-SS) chambers—often called static chambers, absorption chambers, or alkali trap chambers—were the first devices used for measuring the flux of CO_2 from the soil surface (Bornemann, 1920; Lundegårdh, 1921) and were the only devices used very extensively for this purpose before the mid-1980s. Such chambers contain a trap filled with a known amount of alkali solution (or soda lime) that is supported above the soil surface, and they are typically deployed for long periods, often 12 or 24 h.

The amount of CO_2 trapped by the alkali is determined by titration (or by weight change for soda lime) and is usually corrected for CO_2 absorbed by an identical (blank) trap that is handled the same as other traps, but placed in a chamber deployed over a nonemitting surface. The resulting estimate of the soil respiration rate is often smaller than occasionally greater than estimates obtained from nonsteady-state (NSS) and flow-through steady-state (FT-SS) chamber systems, both of which gained rapid acceptance when accurate and portable CO_2 analysers became widely available about two decades ago. As a result, NFT-SS chamber estimates of soil respiration are considered by many investigators to be unreliable and, at best, to be estimates of the relative differences among various sources

The rate of gaseous CO_2 absorption by an alkali trap is proportional to the CO_2 concentration to which it is exposed. In a perfectly designed NFT-SS chamber, the absorption rate is equal to the surface flux, F_c, when headspace CO_2 is equal to the ambient level. In that situation, no change would occur in concentrations of the gas either above or below the soil surface during the period of deployment, F_c would remain equal to the underlying rate of CO_2 production, and the chamber would provide an accurate estimate of soil respiration.

However, achieving and maintaining such a balance is difficult, if not impossible, and the CO_2 concentration of both headspace and subsurface air often rises following chamber deployment until the rate of CO_2 absorption becomes equal to F_c The result is an underestimate of soil respiration—first, because the increase represents CO_2 production not accounted for in the alkali trap, and second, because the elevated concentrations support CO_2 losses by leakage through imperfect chamber seals and by lateral diffusion beneath the chamber walls. In contrast, headspace and subsurface CO_2 concentrations may decline following chamber deployment if F_c is particularly small. The result in this case is an overestimate of soil respiration, because the decline represents absorbed CO_2 not produced during the deployment period, and because the reduced concentrations support CO_2 gain by the chamber system via leakage and subsurface lateral diffusion.

The amount of increase (or decrease) in headspace CO_2 concentration depends not only on the magnitude of the soil respiration rate, but also on the efficiency of CO_2 absorption by the alkali trap, the effective diffusivity and storage coefficient of the gas in underlying soil, and its rate of exchange (leakage) between the chamber system and its surroundings. In a recent exhaustive review of existing literature regarding NFT-SS chambers, Rochette and Hutchinson (2003) concluded that:

- The optimal strength of the alkali solution is 0.5 to 1.0 *M*,
- The alkali trap should have total capacity approximately three times greater than the amount of CO_2 expected to be emitted during the deployment period,
- A 20% ratio of exposed alkali trap area to emitting soil surface area provides good absorption efficiency in many situations, but can be altered when needed to keep headspace CO_2 concentration as close as possible to the ambient level,
- The chamber should be nonvented and should have good seals that minimize CO_2 exchange between the chamber and its surroundings, and
- The deployment period should be at least 12 and preferably 24 h to minimize measurement bias due to the initial nonsteady-state condition, as well as bias due to chamber-induced temperature disturbances that often differ in algebraic sign between day and night (and thus tend to at least partially cancel across a 24-h deployment). Despite the decades of experience and scores of studies summarized by this review, however, many questions remain regarding the accuracy and functioning of NFT-SS chambers, as well as the optimal protocol for their use.

APPROACH

The transfer of gaseous CO_2 from its subsurface point of production to its carbonate form in the alkali trap of a NFT-SS chamber may be viewed as a seven-step process:

- Diffusive transport through air-filled soil pore space to the soil-atmosphere boundary,
- Diffusive transport through the interfacial layer of

still air overlying the soil surface,

- Diffusive and/or convective transport from the top of that interfacial layer to the top of the interfacial layer overlying the alkali surface,
- Diffusive transport through this second interfacial layer of still air,
- Dissolution at the alkali solution surface,
- Chemical reaction with alkali near the solution surface, and
- Migration of dissolved carbonate away from the alkali surface.

There is considerable textbook and empirical evidence that the last three steps should be nonlimiting, and the weak dependence of absorption efficiency on alkali molarity greater than a modest threshold value supports this notion.

Because each of the first four steps is potentially rate-limiting, and all involve gaseous transport, we used a gas diffusion model to investigate their effect on CO_2 absorption efficiency, as well as their interaction with processes controlling CO_2 gain or loss by the chamber system. The model ignores diffusion of dissolved CO_2 and HCO^-_3 through soil water-filled pore space because of its small contribution to total transport. Other details regarding application of this model to chamber systems are available in Healy et al. and Hutchinson et al.

Briefly, we assumed that the soil was bare and had uniform properties with pH of 6.5, total porosity of 0.5, AFP of 0.3, tortuosity computed from the equation of Sallam et al. and effective diffusivity equal to the three-way product of tortuosity, AFP, and the binary molecular diffusion coefficient of CO_2 in air (*D*). Total porosity and AFP are understood to have units of m^3 m^{-3} soil.

Properties for CO_2 were chosen at 20°C and 100 kPa air pressure: $D = 0.160$ cm^2 s^{-1} ambient atmospheric concentration = 375 μmol mol^{-1}, Henry's Law coefficient for dissolution in water = 0.935 and the dissociation constant of dissolved CO_2 (H_2CO_3) = $10^{-6.40.}$ Equilibrium between gas phase and solution phase CO_2 and among dissolved species in soil water and in the alkali solution was assumed to occur instantaneously. The CO_2 was generated by a constant (or 24-h sinusoidal) zero-order source term with magnitude that decreased exponentially (with 10-cm relaxation depth) from a maximum at the soil surface to zero at the impermeable bottom of the simulated domain 50 cm below the surface.

The nonvented chamber headspace (30-cm diametre x 20-cm height) was assumed to be convectively mixed above a 0.5-cm diffusively mixed atmospheric interfacial layer, the depth of which was not altered by deployment of the chamber. Unless otherwise specified, the convection supported mixing equivalent to a tenfold increase in *D*. Chamber walls were assumed to be inserted 5 cm into the soil, except when studying the effect of changes in this parametre. Wall thickness was 0.25 cm except near the bottom, which we assumed was beveled to provide a cutting edge, as is often done to ease insertion and minimize the resulting potential for soil compaction. Finally, the cylindrical alkali container (5-cm internal depth x 0.2-cm walls) was mounted in the horizontal centre of the chamber with its bottom 2 cm above the soil surface. Its internal diametre was 13.4 cm when the ratio of exposed alkali surface area to emitting soil surface area was 0.20, and liquid depth (regardless of trap area) was 0.5 cm. The diffusively mixed atmospheric interfacial layer immediately above the alkali solution surface also had a 0.5-cm depth, and the convection in overlying air surrounded by vertical walls of the trap was assumed to support mixing equivalent to a threefold increase in *D* (compared with a tenfold increase in the bulk headspace).

Controllers of NFT-SS chamber performance that we investigated included the magnitude and diurnal periodicity of the soil respiration rate, the ratio of exposed alkali surface area to emitting soil surface area (hereafter abbreviated as the alkali:chamber area ratio), the efficiency of convective transport from the top of the soil interfacial layer to the top of the alkali interfacial layer, depth of the latter layer, height of the alkali trap above the soil surface, soil AFP, soil pH (which impacts the equilibrium between dissolved CO_2 and

CO^{-}_{3}/HCO^{-}_{3} in soil water), depth of chamber wall insertion into the soil, and the leakiness of chamber seals.

For convenience of the reader, the values assumed for these variables are listed at the top of all figures that report simulation results; the variable under examination in each case is identified by bold type.

Except in the simulations reported in the surface flux prior to and including the moment of chamber deployment was defined as the steady-state CO_2 flux density under the assumed conditions (equal to the depth-integrated rate of CO_2 production), which we designated F_o. Chamber performance was examined in the top graph of, 4, 5, and 6 by normalizing the instantaneous simulated flux into the chamber (F_c) with respect to F_o and plotting the result as a function of time after deployment. Deviations of the resulting trace from a horizontal straight line at $F_c/F_o = 1$ then represent a measure of chamber-induced perturbation of the pre-deployment CO_2 exchange rate.

The bottom graph in these three-panel shows the mean CO_2 concentration of the headspace as a function of time, while the bar graphs represent CO_2 accumulated by the alkali trap after several time periods (expressed as flux density, F_a); like F_c, F_a was normalized with respect to F_o. The solid curves in the top and bottom graphs of these figures identify the level at which the variable under investigation was held constant in other simulations while another was varied to study its influence on NFT-SS chamber performance.

Simulated non-flow-through steady-state chamber performance as a function of time when F_o (the depth-integrated rate of subsurface CO_2 production) followed a sinusoidal pattern with a 24-h period (solid curve in bottom panel graph); other simulation conditions are listed at the top of the figure. Curves with broken line types in the graph represent F_a (the alkali trap's instantaneous CO_2 absorption rate) during four simulated 24-h chamber deployment periods beginning at 0, 6, 12, or 18 h. Values in the centre-panel table are cumulative measurement error for the first 12 h or entire 24 h of each deployment period.

Simulated non-flow-through steady-state chamber performance as a function of time and F_o (the depth-integrated rate of subsurface CO_2 production) under conditions summarized above each three-panel array of graphs.

Chamber sidewalls were inserted (a) 5 cm or (b) 50 cm into the soil. The top and middle graphs show F_c (the surface flux of CO_2) and F_a (the alkali-measured flux of CO_2), respectively, both. normalized with respect to F_o; bottom graphs show mean headspace CO_2 concentration.

STEADY FLOW ENERGY EQUATION

The sketch above shows a piece of equipment such as a boiler, engine, pump, etc. through which fluid steadily flows. The fluid enters the equipment with velocity V_1 at an inlet 1 with area A_1 and leaves with velocity V_2 by an exhaust at 2 with area A_2. The heights of the inlet and exhaust above some reference datum are z_1 and z_2 respectively.

The equipment is enclosed within an imaginary surface called a control surface. Note that, unlike a system boundary, matter can cross a control surface. The control surface encloses a *control volume* and the objective here is to undertake an energy balance for that control volume.

Heat is transferred across the control surface at rate Q and work is being performed at rate by an output shaft. That work is called shaft work.

Consider a system that initially fills the control volume and a small section of the inlet plumbing outside the control volume. A small time $t\delta$later that system has moved so that it now coincides precisely with the control surface at the inlet, but part of the system has left the control volume at the exhaust.

STEADY FLOW PROCESSES

It should be clear from the above that the heat transfer in a steady flow process is the same as in the equivalent non-flow process, but the work w_s differs from the non-flow work because of the 'flow work' terms. For calculations one usually can find the heat transfer as for a non-flow process of the same kind and then find the shaft work w_s using the steady flow energy equation.

Of particular interest is the steady-flow adiabatic process. One group of equipment includes compressors and pumps of all kinds that are driven by a shaft and are used to raise the pressure of a flowing substance. Another group includes reciprocating steam engines and turbines of various kinds, which have an output shaft driving some load and derive their work output by reducing the pressure of the working substance flowing through them. None of these devices has any deliberate heat transfer, and any heat transfer that does occur is trivial. Consequently the process taking place is idealised as adiabatic and steady flow. Furthermore on these devices the potential energy and kinetic energy terms in the steady-flow energy

GAS LAWS

BOYLE'S LAW

Torricelli's experiment did more than just show that air has weight; it also provided a way of creating a vacuum because the space above the column of mercury at the top of a barometre is almost completely empty. (It is free of air or other gases except a negligible amount of mercury vapour.) Torricelli's work with a vacuum soon caught the eye of the British scientist Robert Boyle.

Boyle's most famous experiments with gases dealt with what he called the "spring of air." These experiments were based on the observation that gases are*elastic*. (They return to their original size and shape after being stretched or squeezed.) Boyle studied the elasticity of gases in a J-tube similar to the apparatus shown in the figure below. By adding mercury to the open end of the tube, he trapped a small volume of air in the sealed end.

Boyle studied what happened to the volume of the gas in the sealed end of the tube as he added mercury to the open end. Boyle noticed that the product of the pressure times the volume for any measurement in this table was equal to the product of the pressure times the volume for any other measurement, within experimental error.

$$P_1V_1 = P_2V_2$$

This expression, or its equivalent, is now known as Boyle's Law.

AMONTON'S LAW

Toward the end of the 1600s, the French physicist Guillaume Amontons built a thermometre based on the fact that the pressure of a gas is directly proportional to its temperature. Amontons' law explains why car manufacturers recommend adjusting the pressure of your tires before you start on a trip. The flexing of the tire as you drive inevitably raises the temperature of the air in the tire. When this happens, the pressure of the gas inside the tires increases.

Amontons' law can be demonstrated with the apparatus shown in the figure below, which consists of a pressure gauge connected to a metal sphere of constant volume, which is immersed in solutions that have different temperatures.

In 1779 Joseph Lambert proposed a definition for absolute zero on the temperature scale that was based on the straight-line relationship between the temperature and pressure of a gas shown in the figure above.

He defined absolute zero as the temperature at which the pressure of a gas becomes zero when a plot of pressure versus temperature for a gas is extrapolated. The pressure of a gas approaches zero when the temperature is about -270°C. When more accurate measurements are made, the pressure of a gas extrapolates to zero when the temperature is -273.15°C. Absolute zero on the Celsius scale is therefore -273.15°C.

The relationship between temperature and pressure can be greatly simplified by converting the temperatures from the Celsius to the Kelvin scale.

$$T_K = To_C + 273.15$$

When this is done, a plot of the temperature versus the pressure of a gas gives a straight line that passes through the origin. Any two points along the line therefore fit the following equation. It is important to remember that this equation is only valid if the temperatures are converted from the Celsius to the Kelvin scale before calculations are done.

CHARLES' LAW

Joseph and Etienne Montgolfier used a fire to inflate a spherical balloon about 30 feet in diametre that traveled about a mile and one-half before it came back to earth. News of this remarkable achievement spread throughout France, and Jacques-Alexandre-Cesar Charles immediately tried to duplicate this performance. As a result of his work with balloons, Charles noticed that the volume of a gas is directly proportional to its temperature.

The relationship between the temperature and volume of a gas, which became known as Charles' law, provides an explanation of how hot-air balloons work. Ever since the third century B.C., it has been known that an object floats when it weighs less than the fluid it displaces. If a gas expands when heated, then a given weight of hot air occupies a larger volume than the same weight of cold air. Hot air is therefore less dense than cold air.

Once the air in a balloon gets hot enough, the net weight of the balloon plus this hot air is less than the weight of an equivalent volume of cold air, and the balloon starts to rise. When the gas in the balloon is allowed to cool, the balloon returns to the ground.

Charles' law can be demonstrated with the apparatus shown in the figure below. A 30-mL syringe and a thermometre are inserted through a rubber stopper into a flask that has been cooled to 0°C. The ice bath is then removed and the flask is immersed in a warm-water bath. The gas in the flask expands as it warms, slowly pushing the piston out of the syringe. The total volume of the gas in the system is equal to the volume of the flask plus the volume of the syringe.

GAY-LUSSAC'S LAW

Gay-Lussac began his career in 1801 by very carefully showing the validity of Charles' law for a number of different gases. Gay-Lussac's most important contributions to the study of gases, however, were experiments he performed on the ratio of the volumes of gases involved in a chemical reaction.

Gay-Lussac studied the volume of gases consumed or produced in a chemical reaction because he was interested in the reaction between hydrogen and oxygen to form water. He argued that measurements of the *weights* of hydrogen and oxygen consumed in this reaction could be influenced by the moisture present in the reaction flask, but this moisture would not affect the *volumes* of hydrogen and oxygen gases consumed in the reaction.

Much to his surprise, Gay-Lussac found that 199.89 parts by volume of hydrogen were consumed for every 100 parts by volume of oxygen. Thus, hydrogen and oxygen seemed to combine in a simple 2:1 ratio by volume.

$$\text{Hydrogen} + \text{oxygen} \rightarrow \text{Water}$$

$$2\,\text{Volumes} \quad 1\,\text{Volume}$$

Gay-Lussac found similar whole-number ratios for the reaction between other pairs of gases. The compound we now know as hydrogen chloride (HCl) combined with ammonia (NH_3) in a simple 1:1 ratio by volume:

$$\text{Hydrogen chloride} + \text{ammonia} \rightarrow \text{ammonium chloride}$$

$$1\,\text{Volume} \qquad\qquad 1\,\text{Volume}$$

Carbon monoxide combined with oxygen in a 2:1 ratio by volume:

carbon monoxide,

$$\text{Carbon monoxide} + \text{Oxygen} \rightarrow \text{Carbon dioxide}$$

$$2\,\text{volumes} \qquad\qquad 1\,\text{volume}$$

Gay-Lussac obtained similar results when he analysed the volumes of gases given off when compounds decomposed. Ammonia, for example, decomposed to give three times as much hydrogen by volume as nitrogen:

$$\text{Ammonia} \rightarrow \text{Nitrogen} + \text{Hydrogen}$$

$$1\,\text{volume} \quad 3\,\text{volumes}$$

On 31 December 1808, Gay-Lussac announced his law of combining volumes to a meeting of the Societ Philomatique in Paris. At that time, he summarized the law as follows: Gases combine among themselves in very simple proportions. Today, Gay-Lussac's law is stated as follows: The ratio of the volumes of gases consumed or produced in a chemical reaction is equal to the ratio of simple whole numbers.

AVOGADRO'S HYPOTHESIS

Gay-Lussac's law of combining volumes was announced only a few years after John Dalton proposed his atomic theory. The link between these two ideas was first recognized by the Italian physicist Amadeo Avogadro three years later, in 1811. Avogadro argued that Gay-Lussac's law of combining volumes could be explained by assuming that equal volumes of different gases collected under similar conditions contain the same number of particles.

HCl and NH_3 therefore combine in a 1:1 ratio by volume because one molecule of HCl is consumed for every molecule of NH_3 in this reaction and equal volumes of these gases contain the same number of molecules.

$$NH_3(g) + Hcl(g) \rightarrow NH_4cl(s)$$

Anyone who has blown up a balloon should accept the notion that the volume of a gas is proportional to the number of particles in the gas.

The more air you add to a balloon, the bigger it gets. Unfortunately this example does not test Avogadro's hypothesis that equal volumes of *different gases* contain the same number of particles.

A small hole is drilled through the plunger of a 50-mL plastic syringe. The plunger is then pushed into the syringe and the syringe is sealed with a syringe cap. The plunger is then pulled out of the syringe until the volume reads 50 mL and a nail is inserted through the hole in the plunger so that the plunger is not sucked back into the barrel of the syringe. The "empty" syringe is then weighed, the syringe is filled with 50 mL of a gas, and the syringe is reweighed. The difference between these measurements is the mass of 50 mL of the gas.

The results of experiments with six gases are given in the table below.

Experimental Data for the Mass of 50-mL Samples of Different Gases,

Compound	Mass of 50 mL of Gas (g)	Molecular Weight of Gas	Number of Gas Molecules
H_2	0.005	2.02	1×10^{21}
N_2	0.055	28.01	1.2×10^{21}
O_2	0.061	32.00	1.1×10^{21}
CO_2	0.088	44.01	1.2×10^{21}
C_4H_{10}	0.111	58.12	1.15×10^{21}
CCl_2F_2	0.228	120.91	1.14×10^{21}

The number of molecules in a 50-mL sample of any one of these gases can be calculated from the mass of the sample, the molecular weight of the gas, and the number of molecules in a mole. Consider the following calculation of the number of H_2 molecules in 50 mL of hydrogen gas, for example.

$$0.005\,g\,H_2 \times \frac{1\,mol\,H_2}{2.02\,g\,H_2} \times \frac{6.02 \times 10^{23}\text{ molecules}}{1\,mol\,H_2}$$

$$= 1 \times 10^{21}\ H_2\text{ molecules}$$

The last column in the table above summarizes the results obtained when this calculation is repeated for each gas. The number of significant figures in the answer changes from one calculation to the next. But the number of molecules in each sample is the same, within experimental error. We therefore conclude that equal volumes of different gases collected under the same conditions of temperature and pressure do in fact contain the same number of particles.

IDEAL GAS EQUATION

Gases can described in terms of four variables: pressure

(P), volume (V), temperature (T), and the amount of gas (n). There are five relationships between pairs of these variables in which two of the variables were allowed to cahnge while the other two were held constant.

	P	$\propto$	n	(T and V constant)
Boyle's law:	P	$\propto$	$1/V$	(T and n constant)
Amontons' law:	P	$\propto$	T	(V and n constant)
Charles' law:	V	$\propto$	T	(P and n constant)
Avogadro's hypothesis:	V	$\propto$	n	(P and T constant)

Each of these relationships is a special case of a more general relationship known as the ideal gas equation.

$$PV = nRT$$

In this equation, R is a proportionality constant known as the *ideal gas constant* and T is the absolute temperature. The value of R depends on the units used to express the four variables P, V, n, and T. By convention, most chemists use the following set of units.

P: atmospheres

T: kelvin

V: liters

n: moles

IDEAL GAS CALCULATIONS

The ideal gas equation can be used to predict the value of any one of the variables that describe a gas from known values of the other three. The key to solving ideal gas problems often involves recognizing what is known and deciding how to use this information.

The ideal gas equation can be applied to problems that don't seem to ask for one of the variables in this equation. The ideal gas equation can even be used to solve problems that don't seem to contain enough information.

Gas law problems often ask you to predict what happens when one or more changes are made in the variables that describe the gas. There are two ways of working these problems. A powerful approach is based on the fact that the ideal gas constant is in fact a constant.

We start by solving the ideal gas equation for the ideal gas constant.

$$R = \frac{PV}{nT}$$

We then note that the ratio of *PV*/*nT* at any time must be equal to this ratio at any other time.

$$\frac{P_1V_1}{n_1T_1} = \frac{P_2V_2}{n_2T_2}$$

We then substitute the known values of pressure, temperature, volume, and amount of gas into this equation and solve for the appropriate unknown. This approach has two advantages. First, only one equation has to be remembered. Second, it can be used to handle problems in which more than one variable changes at a time.

DALTON'S LAW OF PARTIAL PRESSURES

The *Handbook of Chemistry and Physics* describes the atmosphere as 78.084% N_2, 20.946% O_2, 0.934% Ar, and 0.033% CO_2 by volume when the water vapour has been removed. What image does this description evoke in your mind? Do you believe that only 20.463% of the room in which you are sitting contains O_2? Or do you believe that the atmosphere in your room is a more or less homogeneous mixture of these gases?

Gases expand to fill their containers. The volume of O_2 in your room is therefore the same as the volume of N_2. (Both gases expand to fill the room.) When we describe the atmosphere as 20.946% O_2 by volume, we mean that the volume of the atmosphere would shrink by 20.946% if the O_2 is removed.

What about the pressure of the different gases in your room? Is the pressure of the O_2 in the atmosphere the same as the pressure of the N_2? We can answer this question by rearranging the ideal gas equation as follows.

$$p = n \times \frac{RT}{V}$$

According to this equation, the pressure of a gas is

proportional to the number of moles of gas, if the temperature and volume are held constant. Because the temperature and volume of the O_2 and N_2 in the atmosphere are the same, the pressure of each gas must be proportional to the number of the moles of the gas. Because there is more N_2 in the atmosphere than O_2, the contribution to the total pressure of the atmosphere from N_2 is larger than the contribution from O_2.

John Dalton was the first to recognize that the total pressure of a mixture of gases is the sum of the contributions of the individual components of the mixture. By convention, the part of the total pressure of a mixture that results from one component is called the partial pressure of that component. Dalton's law of partial pressures states that the total pressure of a mixture of gases is the sum of the partial pressures of the various components.

$$P_T = P_1 + P_2 + P_3 + ...$$

Dalton derived the law of partial pressures from his work on the amount of water vapour that could be absorbed by air at different temperatures. It is therefore fitting that this law is used most often to correct for the amount of water vapour picked up when a gas is collected by displacing water. Suppose, for example, that we want to collect a sample of O_2 prepared by heating potassium chlorate until it decomposes.

$$2Kclo_3(s) \rightarrow 2Kcl(s) + 3o_2(g)$$

Because some of the water in the flask will evaporate during the experiment, the gas that collects in this flask is going to be a mixture of O_2 and water vapour. The total pressure of this gas is the sum of the partial pressures of these two components.

$$P_T = P_{oxygen} + P_{water}$$

The total pressure of this mixture must be equal to atmospheric pressure. (If it was any greater, the gas would push water out of the container. If it was any less, water would be forced into the container.) If we had some way to estimate the partial pressure of the water in this system, we could therefore calculate the partial pressure of the oxygen gas.

By convention, the partial pressure of the gas that collects in a closed container above a liquid is known as the vapour pressure of the liquid.

If we know the temperature at which a gas is collected by displacing water, and we assume that the gas is saturated with water vapour at this temperature, we can calculate the partial pressure of the gas by subtracting the vapour pressure of water from the total pressure of the mixture of gases collected in the experiment. Gases behave differently from the other two commonly studied states of matter, solids and liquids, so we have different methods for treating and understanding how gases behave under certain conditions.

Gases, unlike solids and liquids, have neither fixed volume nor shape. They are molded entirely by the container in which they are held. We have three variables by which we measure gases:pressure, volume, and temperature. Pressure is measured as force per area. The standard SI unit for pressure is the pascal (Pa). However, atmospheres (atm) and several other units are commonly used. The table below shows the conversions between these units.

Table. Units of Pressure

1 pascal (Pa)	1 N^*m^{-2} = 1 $kg^*m^{-1*}s^{-2}$
1 atmosphere (atm)	1.01325^*10^5 Pa
1 atmosphere (atm)	760 torr
1 bar	10^5 Pa

Volume is related between all gases by Avogadro's hypothesis, which states: Equal volumes of gases at the same temperature and pressure contain equal numbers of molecules. From this, we derive the molar volume of a gas (volume/moles of gas). This value, at 1 atm, and 0° C is shown below.

$$vm = \frac{v}{n} = 22.4\,L \text{ at } 0°C \text{ and } 1 \text{ atm}$$

Where:

V_m = molar volume, in liters, the volume that one mole of gas occupies under those conditions

V=volume in liters

n=moles of gas

An equation that chemists call the Ideal Gas Law, shown below, relates the volume, temperature, and pressure of a gas, considering the amount of gas present.

$$PV = nRT$$

Where:

P=pressure in atm

T=temperature in Kelvins

R is the *molar gas constant,* where R=0.082058 L atm $mol^{-1} K^{-1}$.

The Ideal Gas Law assumes several factors about the molecules of gas. The volume of the molecules is considered negligible compared to the volume of the container in which they are held. We also assume that gas molecules move randomly, and collide in completely elastic collisions. Attractive and repulsive forces between the molecules are therefore considered negligible.

Example

Problem: A gas exerts a pressure of 0.892 atm in a 5.00 L container at 15°C. The density of the gas is 1.22 g/L. What is the molecular mass of the gas?

Answer:

PV = nRT

T = 273 + 15 = 228

(0.892)(5.00) = n(.0821)(288)

n = 0.189 mol

$$\frac{0.189\,\text{mol}}{5.00\text{ L}} \times \frac{\text{grams}}{1\,\text{mol}} = 1.22\,\text{g/L}$$

x = Molecular Weight = 32.3 g/mol

We can also use the Ideal Gas Law to quantitatively determine how changing the pressure, temperature, volume, and number of moles of substance affects the system. Because the gas constant, R, is the same for all gases in any situation, if you solve for R in the Ideal Gas Law and then set two Gas Laws equal to one another, you have the Combined Gas Law:

$$\frac{P_1 v_1}{n_1 T_1} = \frac{P_2 V_2}{n_2 T_2}$$

Where:

values with a subscript of "1" refer to initial conditions
values with a subscript of "2" refer to final conditions.

If you know the initial conditions of a system and want to determine the new pressure after you increase the volume while keeping the numbers of moles and the temperature the same, plug in all of the values you know and then simply solve for the unknown value.

Example

Problem: A 25.0 mL sample of gas is enclosed in a flask at 22°C. If the flask was placed in an ice bath at 0°C, what would the new gas volume be if the pressure is held constant?

Answer: Because the pressure and the number of moles are held constant, we do not need to represent them in the equation because their values will cancel. So the combined gas law equation becomes:

$$\frac{V_1}{T_1} = \frac{V_2}{T_2}$$

$$V_2 = 23.1 \text{ mL}$$

We can apply the Ideal Gas Law to solve several problems. Thus far, we have considered only gases of one substance, pure gases. We also understand what happens when several substances are mixed in one container. According to Dalton's law of partial pressures, we know that the total pressure exerted on a container by several different gases, is equal to the sum of the pressures exerted on the container by each gas.

$P_t = P_1 + P_2 + P_3 + \ldots$

Where:

P_t=total pressure
P_1=partial pressure of gas "1"
P_2=partial pressure of gas "2"
and so on

Using the Ideal Gas Law, and comparing the pressure of one gas to the total pressure, we solve for the mole fraction.

$$\frac{P_1}{P_t}=\frac{n_2RT/V}{n_tRT/V}=\frac{n_1}{n_t}=X_1$$

Where:

X_1 = mole fraction of gas "1"

And discover that the partial pressure of each the gas in the mixture is equal to the total pressure multiplied by the mole fraction.

$$P_1=\frac{n_1}{n_t}P_t=X_1P_t$$

Chapter 2

Fundamental Laws

DEFINITION

The Laws of Thermodynamics are considered part of the bedrock of physics (and among physicists, about as immobile as bedrock). But there are two problems.

PROBLEMS

First Problem

The first is essentially a clarification. The Laws of Physics (including specifically the Laws of Thermodynamics) are, in reality, *not* laws of nature, but instead man-made theories or informed guesses of how man believes nature works and how she apparently limits herself.

They are a theory, and no amount of propaganda or public relations will convert them magically into fundamental laws which cannot be breached. It takes, as a matter of fact, only *one* contrary example to shake the most basic bedrock of theory. And the current problem is that we're all living in an earthquake zone!

The most noteworthy example of physics undergoing a forced revision of physical laws concerns the Law of Conservation of Energy, and prior to the time of Einstein's famous $E=mc^2$ equation, the Law of Conservation of Mass (aka matter, material). Both of these "laws" stated that in a closed system, the total amount of Energy (or Mass) could not change. Period. This was the equivalent of an economic Zero-Sum Game, in which the debits and credits must balance out to zero. Each of the two conservation laws (energy and mass) were independent, prior to Einstein. But with the recognition that mass was a form of energy, the situation changed. Suddenly, the Law of Conservation of Mass went out-of-fashion, and was very quietly left to die alone and in callous disrepute — like an unwanted relative down on his luck. The resulting Law of Conservation of Energy — as modified by Einstein and including mass as a form of energy — was once again sacrosanct. Unfortunately, the lesson of the "fundamental law" needing a fundamental rewrite was lost on several generations of scientists, economists, and the world at large.

The revised Law of the Conservation of Energy, also became known in physics as the First Law of Thermodynamics. It was as if a modification was needed in order to keep the dignity of the laws intact, and simultaneously, try to forget about the unfortunate incident with mass.

Second Problem

The second problem with the Law of Conservation of Energy (aka the First Law of Thermodynamics) is that one of the Assumptions on which it is based is often neglected in the mathematical treatment of the law and the results which are derived from it. No rational physicist would argue that fundamental to the Law of Conservation of Energy is the restraint or assumption that we are dealing with *a closed system!* If the system is not closed, then the law is not strictly applicable, and thus there is no violation of the law.

The same need for a closed system exists in the Second Law of Thermodynamics which basically states that the "order of a system must always decrease" — or alternatively, Entropy, physics' measure of disorder, must always increase.

The difficulty is that most systems — even when they are believed to be a closed system — are quite the contrary.

Physicists who assume they can achieve a closed system in their experiments are simply wrong. The ideal, closed system is much harder to achieve than one might imagine. In fact, Quantum Physics — notably Heisenberg's Uncertainty Principle — states that the very act of observation of an experiment is an intrusion into the system, and effectively alters the experiment.

Furthermore, if as the world's philosophical and spiritual traditions claim (i.e. theorize) that we live in a connected universe, then there are *no closed systems* except for one of universal size. (And that might be invalid as well, if parallel or Multiple Universes exist!)

The Fifth Element and Zero-Point Energy suggest that everything in the universe is in fact connected. On the one hand, Mach's Principle claims inertia is due to the interaction of all masses in the universe, while more recently it has been demonstrated mathematically in the arena of Zero-Point Energy, that all electric charges in the universe interact. The Fifth Element theory goes on to show that there are no limits to the energy that might be conveyed from one entity to another. In all respects there are no closed systems, only approximations.

Back at the physics ranch, the Laws of Thermodynamics must then be viewed as useful tools in which we can accomplish all manner of conjecture and ultimately achieve an effective technology. But these Laws are*always approximations,* and can *in principle never be used to eliminate alternative possible scenarios.*

The great danger is that we forget the limitations of the laws, and assume them to be without exception. However, the laws are correct only *in a closed system,* but inasmuch as there are *no closed systems in the universe,* the laws cannot always be used to disprove other more radical theories.

The laws of thermodynamics describe some of the fundamental truths of thermodynamics observed in our Universe. Understanding these laws is important to students of Physical Geography because many of the processes studied involve the flow of energy.

FIRST LAW OF THERMODYNAMICS

The first law of thermodynamics is often called the *Law*

of Conservation of Energy. This law suggests that energy can be transferred from one system to another in many forms. Also, it can not be *created* or *destroyed*. Thus, the total amount of energy available in the Universe is constant. Einstein's famous equation (written below) describes the relationship between energy and matter:

$$E = mc^2$$

In the equation above, energy (E) is equal to matter (m) times the square of a constant (c). Einstein suggested that energy and matter are interchangeable. His equation also suggests that the quantity of energy and matter in the Universe is fixed.

YOU SHOULD KNOW:

- Know the first Law of Thermodynamics
- Use $Q = U + W$;
- Use $W = pV$ at a constant pressure;
- Know about non-flow processes;
- Know about Isothermal and adiabatic changes, constant pressure and constant volume changes;
- Use $pV = nRT$
- Use pV = constant
- Apply the first law of thermodynamics to these processes.

Thermodynamics is the study of heat flows and how they can be put to work. Engines work by converting heat energy into movement energy, which can then do useful jobs of work for us. We need to look at a couple of key words. A system is the object of interest whose behaviour we are monitoring in relation to its surroundings. A flask containing gas is a system; the water bath in which the flask is placed is its surroundings.

The diagram below helps to show the idea:

The Laws of Thermodynamics were the results of work by nineteenth century physicists. Ironically the Second Law came before the First Law. Then a more fundamental law, the Zeroth Law was worked out.

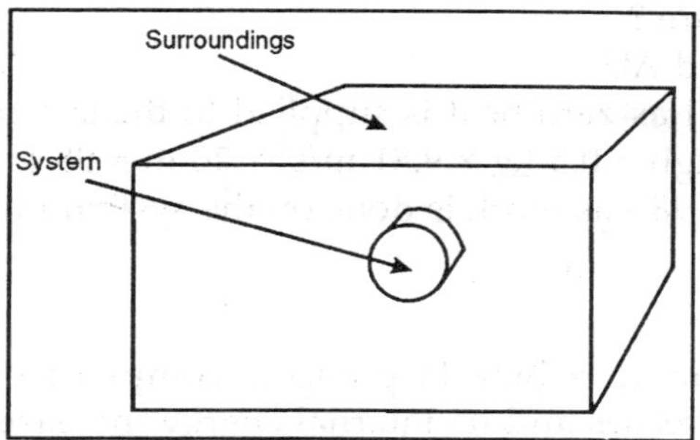

In words the First Law of Thermodynamics is:

The change in internal energy of a system is equal to the sum of energy entering the system through heating and energy entering the system through work done on it.

SOLVED PROBLEMS

Question 1 What is internal energy?

Answer

Internal energy is a measure of the vibration of molecules in a material. In a monatomic gas, it is a measure of kinetic energy of the atoms of gas. In molecular gases, the internal energy is representative of the sum of kinetic energy of the molecules and the vibration of the bonds.

We can write the first law in code:

$$\Delta Q = \Delta U + \Delta W$$

[Q = heat entering the system; U = increase in internal energy; ?W - work done by the system]

The diagram here explains the idea:

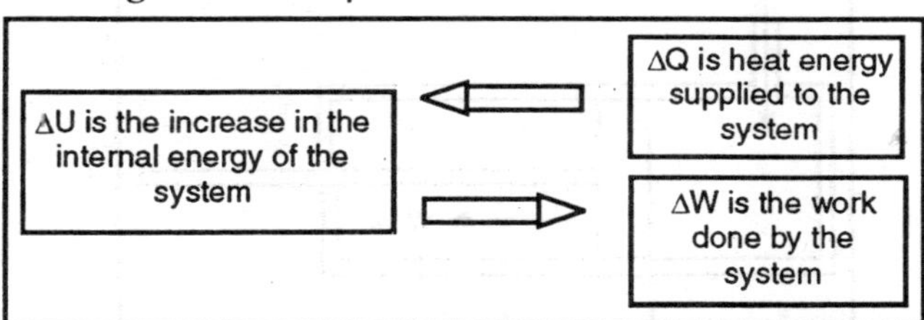

A lump of lead of mass 0.50 kg is dropped from a height of 20 m onto a hard surface. It does not bounce but remains at rest.

What are ?

ΔW, and ΔU.

ΔQ = 0 J as zero heat is supplied to the system

$\Delta U = mgh = 0.5\ kg \times 9.81\ m/s^2 \times 20\ m = 98\ J$

ΔW = – 98 J as work is done on the system rather than by the system

Question 2

Some air in a bicycle pump is compressed so that its volume decreases and its internal energy increases. If 25 J of work are done by the person compressing the air, and if 20 J of thermal energy leave the gas through the walls of the pump, what is the increase in the internal energy of the air?

Answer

Energy supplied = increase in internal energy + work done by system,

25 J = ΔU + 20 J

ΔU = 5 J

If we compress a gas in a bicycle pump, we find it gets hot. Then we let the pump cool down.

Question 3

What happens if we release the pump?

Answer

The pump will spring back to where it was. The gas will cool down, having gained its energy from the surroundings. Consider a cylinder of area *A*. A fluid is admitted at a constant pressure, *p*.

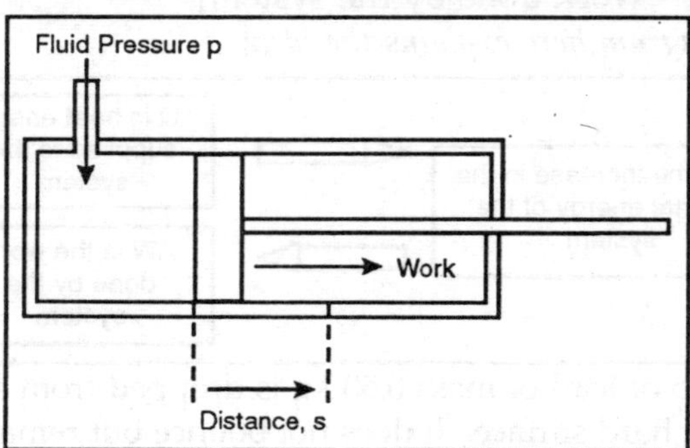

We know that:

- Pressure = force ÷ area
- Work done = force x distance moved

Therefore,

- Force = pressure x area
- Work done = pressure x area x distance moved
- Area x distance moved = change in volume

So we can write:

Work done = pressure x change in volume

BEHAVIOUR OF GASES

ISOTHERMAL CHANGES

In this chapter we saw that the behaviour of ideal gases is governed by the equation:

$$pV = nRT$$

If we keep the temperature the same, we can say that pV = constant, which you may remember as Boyle's Law. Keeping the temperature the same is called an isothermal compression or expansion. We can sketch a graph:

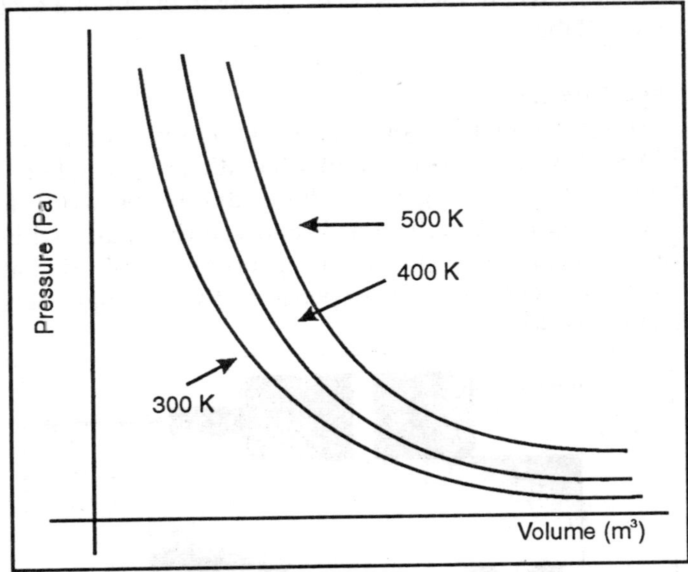

Each of the lines is called an isothermal, because the

temperature is kept the same. We can make the compression and expansion of gases very nearly isothermal by pressing down on a bicycle pump very slowly, so that any heat generated can flow out very slowly. Similarly we can allow the gas to expand very slowly so that the heat flow in in very slow.

For all isothermal processes:

- pV = constant and p1V1 = p2V2;
- ?U = 0 because the internal energy is dependent on temperature, which does not change.
- ?Q = ?W. If the gas expands to do work ?W, and amount of heat ?Q must be supplied

The process is a reversible isothermal change if the piston of the pump is allowed to expand after compression and follows exactly the same line on the graph that it did when being compressed, and ends up in exactly the same place as when it started.

Adiabatic Changes

A change where there is no heat flow in or out of a system is called adiabatic.

$$\Delta Q = 0,$$

therefore $\Delta W = \Delta U$

If you push the plunger of a bicycle pump in very rapidly and block off the end, you get an adiabatic process where the temperature rise of the gas is entirely due to the work done in compressing the gas. If a gas is allowed to expand without any heat energy being put in, the process is still adiabatic. The expansion occurs at the expense of the internal energy. The gas cools down.

An example of this is a little rocket that can be made with a very small cylinder of carbon dioxide at high pressure. This is shown in the diagram above.

The heat flow through the side is negligible compared with the energy loss that causes the drop in temperature as a result of the expansion of the gas. We can use the gas laws to work out the temperature loss.

A useful equation is:

$$\frac{p_1 V_1}{T_1} = \frac{p_2 V_2}{T_2}$$

Isovolumetric Processes

Isovolumetric processes occur at constant volume. We can show this on a graph that displays the isothermals as we have above.

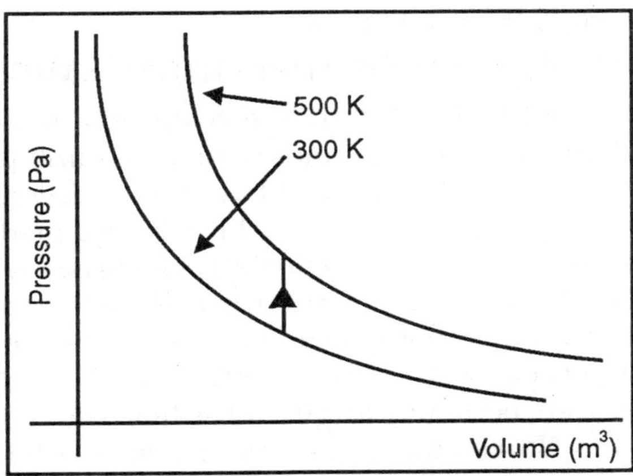

- The process occurs at constant volume
- p1/T1 = p2/T2
- Since there is no change in volume, no work is done, so all heat entering the gas becomes internal energy.

Isobaric Processes

These happen at a constant pressure.

The graph shows the idea:

- The process occurs at constant pressure.
- V1/T1 = V2/T2.

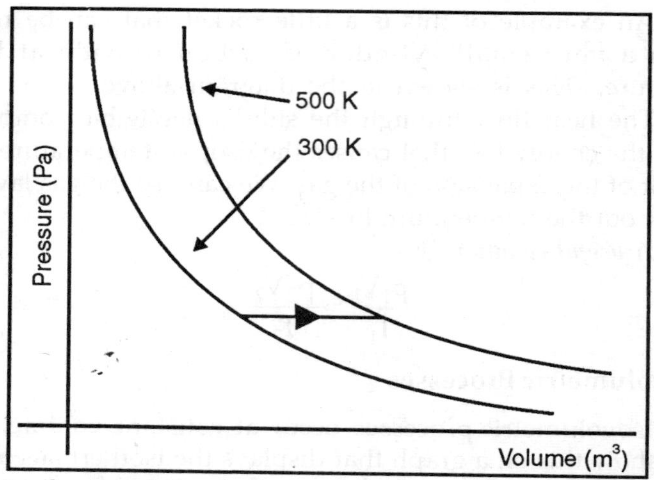

- Some of the heat is used to increase the internal energy, the rest to do work.

SECOND LAW OF THERMODYNAMICS

Heat cannot be transfer from a colder to a hotter body. As a result of this fact of thermodynamics, natural processes that involve energy transfer must have one direction, and all natural processes are irreversible. This law also predicts that the entropy of an isolated system always increases with time. Entropy is the measure of the disorder or randomness of energy and matter in a system. Because of the second law of thermodynamics both energy and matter in the Universe are becoming less useful as time goes on. Perfect order in the Universe occurred the instance after the Big Bang when energy and matter and all of the forces of the Universe were unified.

THIRD LAW OF THERMODYNAMICS

The third law of thermodynamics states that if all the thermal motion of molecules (kinetic energy) could be removed, a state called absolute zero would occur. Absolute zero results in a temperature of 0 Kelvins or -273.15° Celsius.

Absolute Zero = 0 Kelvins = -273.15° Celsius

The Universe will attain absolute zero when all energy

and matter is randomly distributed across space. The current temperature of empty space in the Universe is about 2.7 Kelvins.

CONSEQUENCES OF THE LAWS

The laws of thermodynamics tend to be fairly easy to state and understand... so much so that it's easy to underestimate the impact they have.

Among other things, they put constraints on how energy can be used in the universe. It would be very hard to over-emphasize how significant this concept is. The consequences of the laws of thermodynamics touch on almost every aspect of scientific inquiry in some way.

UNDERSTANDING LAWS OF THERMODYNAMICS

To understand the laws of thermodynamics, it's essential to understand some other thermodynamics concepts that relate to them.

- Thermodynamics Overview - an overview of the basic principles of the field of thermodynamics
- Heat Energy - a basic definition of heat energy
- Temperature - a basic definition of temperature
- Introduction to Heat Transfer - an explanation of various heat transfer methods.
- Thermodynamic Processes - the laws of thermodynamics mostly apply to thermodynamic processes, when a thermodynamic system goes through some sort of energetic transfer.

DEVELOPMENT OF LAWS OF THERMODYNAMICS

The study of heat as a distinct form of energy began in approximately 1798 when Sir Benjamin Thompson (also known as Count Rumford), a British military engineer, noticed that heat could be generated in proportion to the amount of work done... a fundamental concept which would ultimately become a consequence of the first law of thermodynamics.

French physicist Sadi Carnot first formulated a basic principle of thermodynamics in 1824. The principles which Carnot used to define his *Carnot cycle* heat engine would ultimately translate into the second law of thermodynamics by the German physicist Rudolf Clausius, who is also frequently credited with the formulation of the first law of thermodynamics. Part of the reason for the rapid development of thermodynamics in the nineteenth century was the need to develop efficient steam engines during the industrial revolution.

KINETIC THEORY AND THE LAWS OF THERMODYNAMICS

The laws of thermodynamics do not particularly concern themselves with the specific how and why of heat transfer, which makes sense for laws that were formulated before atomic theory was fully adopted. They deal with the sum total of energy and heat transitions within a system, and do not take into account the specific nature of heat transference on the atomic or molecular level.

ZEROTH LAW OF THERMODYNAMICS

Two systems in thermal equilibrium with a third system are in thermal equilibrium to each other. This zeroeth law is sort of a transitive property of thermal equilibrium. The transitive property of mathematics says that if,

A = B and B = C,

then A = C.

The same is true of thermodynamic systems that are in thermal equilibrium. One consequence of the zeroeth law is the idea that measuring temperature has any meaning whatsoever. In order to measure a temperature, thermal equilibrium much be reached between the thermometre as a whole, the mercury inside the thermometre, and the substance being measured. This, in turn, results in being able to accurately tell what the temperature of the substance is.

This law was understood without being explicitly stated through much of the history of thermodynamics study, and it was only realized that it was a law in its own right at the beginning of the 20th century. It was British physicist Ralph H. Fowler who first coined the term "zeroeth law," based on a belief that it was more fundamental even than the other laws.

INTERNAL ENERGY

DEFINITION

Internal energy is defined as the energy associated with the random, disordered motion of molecules. It is separated in scale from the macroscopic ordered energy associated with moving objects; it refers to the invisible microscopic energy on the atomic and molecular scale. For example, a room temperature glass of water sitting on a table has no apparent energy, either potential or kinetic . But on the microscopic scale it is a seething mass of high speed molecules traveling at hundreds of metres per second. If the water were tossed across the room, this microscopic energy would not necessarily be changed when we superimpose an ordered large scale motion on the water as a whole.

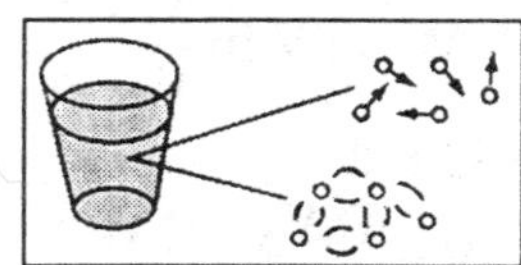

U is the most common symbol used for internal energy.

Related energy quantities which are particularly useful in chemical thermodynamics.

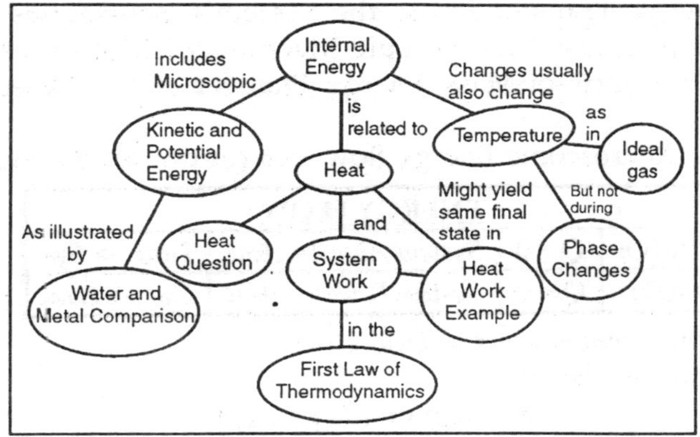

WORK - W, HEAT - Q, AND INTERNAL ENERGY – U WORK W:

Useful Energy Transfered across the System's Bound aries, capable of producing Macroscopic-Mechanical Motion of a the system's Centre-of-Mass.

W = Work done by (or on) one system on another system

ENERGY FLOW			
OUT:	**W > *0***	*System Does External Work*	*Sys --> Work*
INTO:	**W < *0***	*Work Done on the System*	*Work --> Sys*

Work done by a Gas:

$$w = \int_o^f P\,dV$$

Constant Volume Process:

$$W=0$$

Constant Pressure Process:

$$W = P\Delta V$$

Constant Temperature Process:

$$W = PV\,\text{In}\frac{V_f}{V_0}$$

Adiabatic Process Q = 0

$$W = \frac{P_0V_0 - P_fV_f}{\gamma - 1}$$

HEAT Q

Energy Transfer across the System's Boundaries that cannot produce Macroscopic-Mechanical Motion of the system's Centre-of-Mass. Energy Transfer at the Molecular Level .

Q = Microscopic Energy flow into (or out of) the System

ENERGY FLOW			
INTO:	Q > 0	*System Absorbs Heat*	*Heat --> Sys*
OUT:	Q < 0	*System Releases Heat*	*Sys --> Heat*

Some common types of Heat Lost:

Solids or Liquids:

$$Q = mC\Delta T \pm mL_f \pm mL$$

Gas- Constant Pressure Process:

$$Q = mC_p\Delta T$$

Gas - Constant Volume Process:

$$Q = mC\Delta T$$

Gas - Constant Temperature Process:

$$Q = PV\,In\frac{V_f}{V_0}$$

Gas - Adiabatic Process:

$$Q = 0$$

INTERNAL ENERGY U

Energy Stored in a System at the Molecular Level. The System's Thermal Energy -the Kinetic Energy of the atoms due to their random motion relative to the Centre of Mass plus the binding energy (Potential Energy) that holds the atoms together.

U = Microscopic Energy contained in the System

First Law Of Thermodynamics:

Any Change in the Internal Energy of a System U is due to either the Heat Flow into/out-of the System or due to Work Done by/on the System provided the system's centre-of-mass energy does not change.

$$\Delta U = Q - W$$

$$Q = Q_{in} - Q_{out} \quad W = W_{out} - W_{in}$$

- It is important to observe that the ? in ?U is absolutely necessary because both work W and heat Q represent a transfer of energy where as the internal energy U is a quantity of energy that a system contains.
- Another way to state his difference is that U is a state variable were as Q and W are not state variables. What this means is that if you take the system from one state to another state by two different processes, ?U will be the same independent of the path taken but not Q or W.

- The quantities Q_{in}, Q_{out}, W_{in}, and W_{out} are all taken to be positive quantities where as Q and W can be either positive or negative.

APPLICATION OF FIRST LAW

ISOMETRIC PROCESS

It occurs at constant V. Thus, W = 0. From the 1st rule of thermodynamics,

"U = Qv = n cv "T

Quantity of heat supplied to a system at constant V is thus equal to the increase in the internal E. of the system.

ISOBARIC PROCESS

It occurs at constant P. e.g. water boiled and vapourized in a steam engine.

W = +" P dV = P +" dV = P (Vf – Vi) ; Q = m Lv.

From the 1st rule of thermodynamics,

W = m Lv – p (Vf – Vi)

Where Vf = volume of vapour, Vi = volume of liquid.

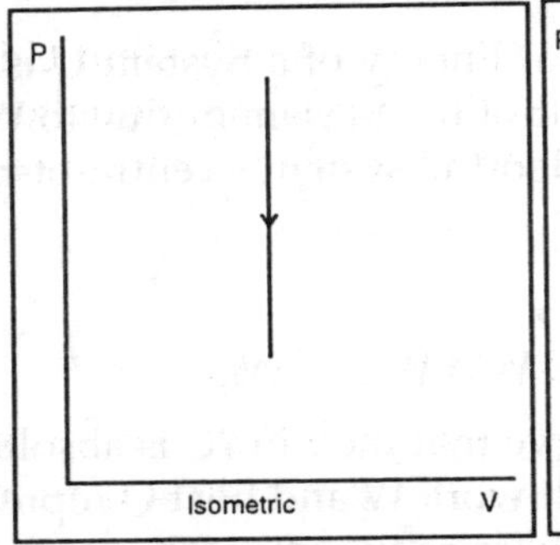

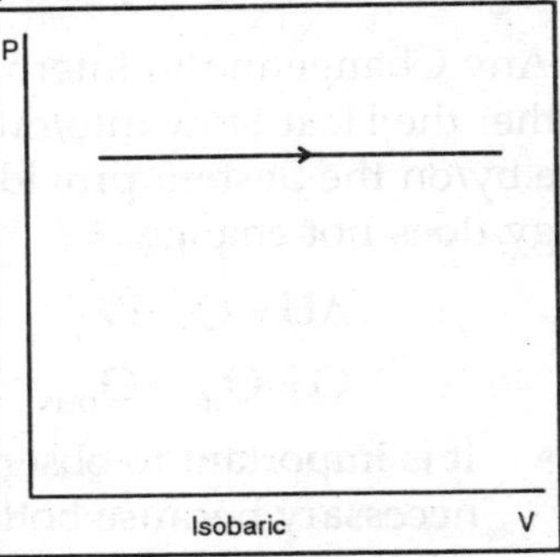

CYCLIC PROCESS

In which final state and initial state are the same. Since internal E. is a state function, internal E. doesn't change during cyclic process.

Thus, "U = 0.

From the 1st rule of thermodynamics,

Q = W

Thus, in a cyclic process, the amount of heat gained or lost through the cycle is equal to the amount of work done = area enclosed by the cycle.

ISOTHERMAL PROCESS

It occurs at constant T. e.g. in thermostats. Thus dT = 0.

P = n R T/V

WT = +" (n R T/V) dV

= n R T ln (Vf/Vi).

(Vf/Vi) = (pi/pf).?

Since T is constant,

Pi Vi = Pf Vf

WT = n R T ln (Vf/Vi)

= n R T ln (pi/pf)

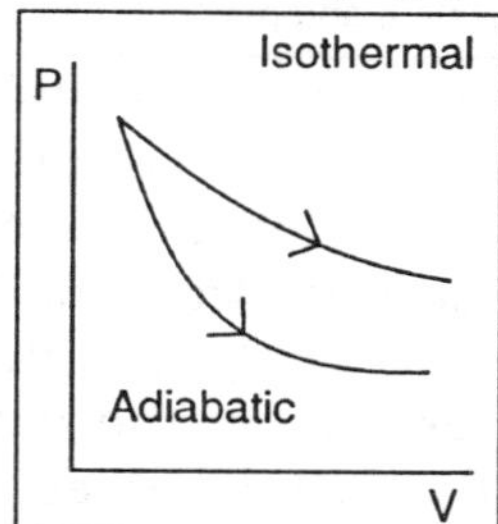

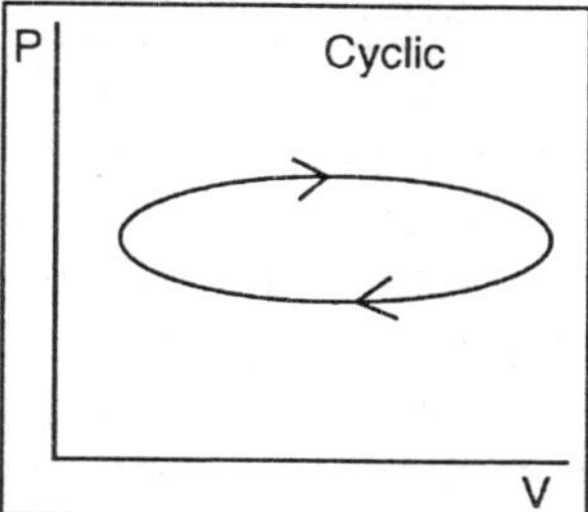

ADIABATIC PROCESS

It occurs so that no heat flows into or out of the system. E.g. by sealing the system off from its surroundings with insulating material or by performing the process quickly (because flow of heat is slow).

Thus Q = 0.

From 1st rule of thermodynamics,

"U = Uf – Ui = – W

Thus, if work is done on the system, internal E. increases by exactly the amount of work done on the system. If work is done by the system (gas expand), internal E. decreases by exactly the amount of external work it does.

Example: Applications of the First Law to motivate the use of a property called enthalpy". Adiabatic, steady, throttling of a gas (flow through a valve or other restriction).

Shows the configuration of interest. We wish to know the relation between properties upstream of the valve, denoted by "1" and those downstream, denoted by "2".

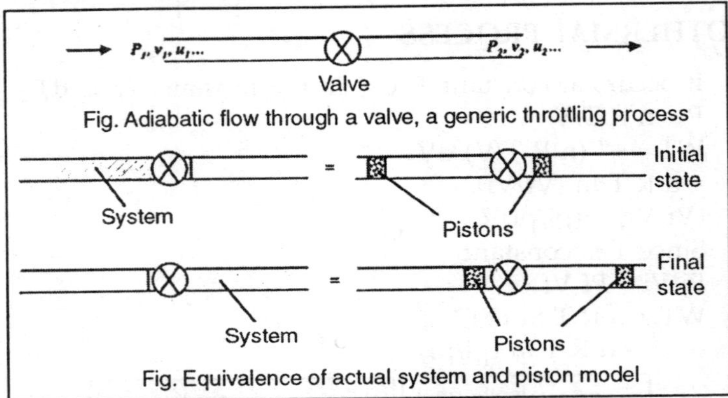

Fig. Adiabatic flow through a valve, a generic throttling process

Fig. Equivalence of actual system and piston model

To analyse this situation, we can define the system (choosing the appropriate system is often a critical element in effective problem solving) as a unit mass of gas in the following two states. Initially the gas is upstream of the valve and just through the valve. In the final state the gas is downstream of the valve plus just before the valve. The figures show the actual configuration just described. In terms of the system behaviour, however, we could replace the fluid external to the system by pistons which exert the same pressure that the external fluid exerts, as indicated schematically.

The process is adiabatic, with changes in potential energy and kinetic energy assumed to be negligible. The first law for the system is therefore

$$\Delta U = -W$$

The work done by the system is,

$$W = P_2V_2 - P_1V_1$$

Use of the first law leads to,

$$U_2 + P_2V_2 = U_1 + P_1V_1$$

In words, the initial and final states of the system have the same value of the quantity U+PV. For the case examined, since we are dealing with a unit mass, the initial and final states of the system have the same value of $u+Pv$.

We define this quantity as the "enthalpy," usually denoted by H,

$$H=U+PV$$

In terms of the specific quantities, the enthalpy per unit mass is,

$$h=u+Pv=u+P/\rho$$

It is a function of the state of the system. H has units of Joules, and h has units of Joules per kilogram.

The utility and physical significance of enthalpy will become clearer as we work with more flow problems. For now, you may wish to think of it as follows When you evaluate the energy of an object of volume , you have to remember that the object had to push the surroundings out of the way to make room for itself. With pressure p on the object, the work required to make a place for itself is pV. This is so with any object or system, and this work may not be negligible. (The force of one atmosphere pressure on one square metre is equivalent to the force of a mass of about 10 tons .) Thus the total energy of a body is its internal energy plus the extra energy it is credited with by having a volume V at pressure p. We call this total energy the enthalpy, H.

QUASI-STATIC EXPANSION OF A GAS

Consider a quasi-static process of constant pressure expansion. We can write the first law in terms of the states before and after the expansion as,

$$Q=(U_2-U_1)+W$$

and writing the work in terms of system properties,

$=(U_2-U_1)+p(V_2-V_1)$

Since $p_1=p_2=p$

By grouping terms we can write the heat input in terms of the enthalpy change of the system:

$Q=(U_2+pV_2)-(U_1+pV_1)$

$= H_2-H_1$

TRANSIENT FILLING OF A TANK

Another example of a flow process, this time for an

unsteady flow, is the transient process of filling a tank, initially evacuated, from a surrounding atmosphere, which is at a pressure P_0 and a temperature T_0. The configuration is shown in Figure.

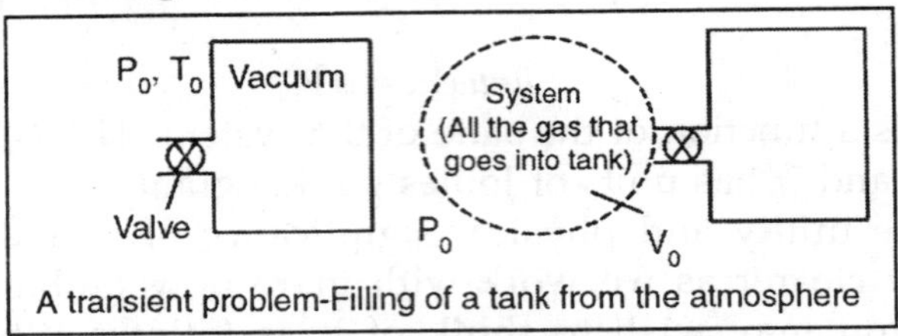

Fig. A Transient Problem—Filling of a Tank from the Atmosphere

At a given time, the valve at the tank inlet is opened and the outside air rushes in. The inflow stops when the pressure inside is equal to the pressure outside. The tank is insulated, so there is no heat transfer to the atmosphere. What is the final temperature of the gas in the tank?

This time we take the system to be all the gas that enters the tank. The initial state has the system completely outside the tank, and the final state has the system completely inside the tank.

The kinetic energy initially and in the final state is negligible, as is the change in potential energy, so the first law again takes the form,

$$\Delta U = -W$$

Work is done *on* the system, of magnitude P_0V_0, where V_0 is the initial volume of the system, so

$$\Delta U = P_0V_0$$

In terms of quantities per unit mass,

($\Delta U = m\Delta u,\ V_0 = mv_0$),

where m is the mass of the system),

$$\Delta u = u_{final} - u_i = P_{0v0}$$

The final value of the internal energy is

$$u_{final} = u_i + P_{0v0}$$
$$= h_i = h_0$$

For a perfect gas with constant specific heats,

$$u = c_v T, h = c_p T,$$

$$c_v T_{final} = c_p T_0$$

$$T_{final} = \frac{c_p}{c_v} T_0 = \gamma T_0$$

The final temperature is thus roughly 200°F hotter than the outside air. It may be helpful to recap what we used to solve this problem.

There were basically four steps:

1. Definition of the system
2. Use of the first law
3. Equating the work to a " PdV" term
4. Assuming the fluid to be a perfect gas with constant specific heats.

A message that can be taken from both of these examples (as well as from a large number of other more complex situations, is that the quantity occurs naturally in problems of fluid flow. Because the combination appears so frequently, it is not only defined but also tabulated as a function of temperature and pressure for a number of working fluids.

FIRST LAW IN TERMS OF ENTHALPY

We start with the first law in differential form and substitute pdV for dW by assuming a quasi-static or reversible process:

$$dU = \delta Q - \delta W$$

(True for any process, neglecting ΔKE and ΔPE) $dU = \delta Q - pdV$ (True for any quasi–static process, no ΔKE OR ΔPE).

The definition of enthalpy,

$$H=U+pV,$$

can be differentiated (applying the chain rule to the pV term) to produce,

$$dH=dU+pdV+Vdp$$

Substituting the dU above for the dU in the First Law, we obtain,

$$dH = \delta Q - \delta W + pdV + Vdp$$

(Valid for any process)

$$dH = \delta Q + Vdp$$

(Valid for any quasi–static process)

HEATING AND EXPANSION OF GASES

EXPANSION OF GAS DUE TO HEATING

This is a differential instrument: the slug of water or mercury in the horizontal section of tubing would be pushed in one direction or the other, depending on the rate at which the air in the two bulbs is expanded by electrical heating. Its ancestor is Kinnersley's Thermometre.

The heating is accomplished by thermocouples in the two bulbs; the manufacturer called this a "Peltier Effect Apparatus." The catalogue notes that "a 2-Ampere flow of current for 10 seconds will produce a 10 to 12 mm movement of a water index in the horizontal connecting tube.

The wooden stand is cut out behind the horizontal tubing, allowing the apparatus to be used with a light source to throw an enlarged image on the screen for the entire class to see.

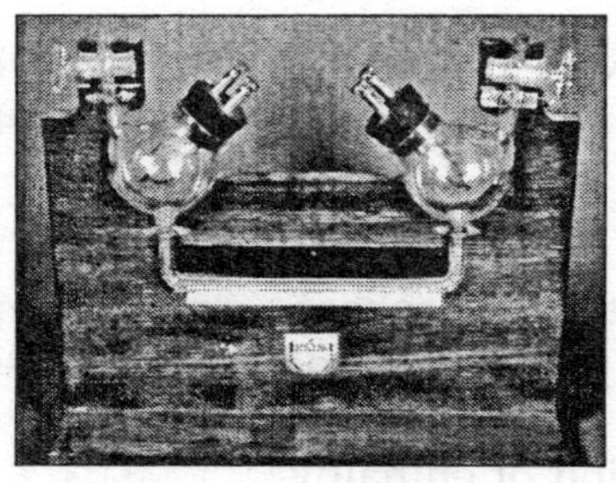

HEATING POWER OF GAS

His is a specialized form of Junkers Calorimetre used to measure the energy obtained by burning gas. The cut below, from the 1917 instruction manual from the American Metre Company of New York, shows the operation of the calorimetre.

The rate at which the gas is burned inside the calorimetre is measured by the wet test metre in the lower picture on the left. Water flows into the jacket around the combustion chamber at a steady rate, controlled by the constant head device at the top.

The mass of the water passing through the apparatus as a function of time is determined with the bucket and scale on the right. The temperature of the incoming water and that of the overflow water is measured by the two thermometres.

FIRST LAW APPLIED TO STEADY FLOW PROCESSES

COMPRESSION-IGNITION) ENGINE

The Air Standard Diesel cycle is the ideal cycle for Compression-Ignition (CI) reciprocating engines, first proposed by Rudolph Diesel over 100 years ago.The four-stroke diesel engine is usually used in motor vehicle systems, whereas larger marine systems usually use the two-stroke diesel cycle.

The actual CI cycle is extremely complex, thus in initial analysis we use an ideal "air-standard" assumption, in which the working fluid is a fixed mass of air undergoing the complete cycle which is treated throughout as an ideal gas. All processes are ideal, combustion is replaced by heat addition to the air, and exhaust is replaced by a heat rejection process which restores the air to the initial state.

ADIABATIC PROCESS OF AN IDEAL GAS (Q = 0)

The analysis results in the following three general forms representing an adiabatic process:

$$Tv^{k-1} = \text{const} \quad TP^{(1-k)/k} = \text{const} \quad Pv^{k} = \text{const}$$

where k is the ratio of heat capacities and has a nominal value of 1.4 at 300K for air.

Process 1-2 is the adiabatic compression process. Thus the temperature of the air increases during the compression process, and with a large compression ratio (usually > 16:1) it will reach the ignition temperature of the injected fuel. Thus given the conditions at state 1 and the compression ratio of the engine, in order to determine the pressure and temperature at state 2 (at the end of the adiabatic compression process) we have:

$$\left(\frac{P_2}{P_1}\right) = \left(\frac{V_1}{V_2}\right)^k = r^k \qquad \left[r = \frac{V_1}{V_2} \Rightarrow \text{Compression Ratio}\right]$$

$$\left(\frac{T_2}{T_1}\right) = \left(\frac{V_1}{V_2}\right)^{k-1} = r^{k-1}$$

Work $W_{1\text{-}2}$ required to compress the gas is shown as the area under the *P-V* curve, and is evaluated as follows.

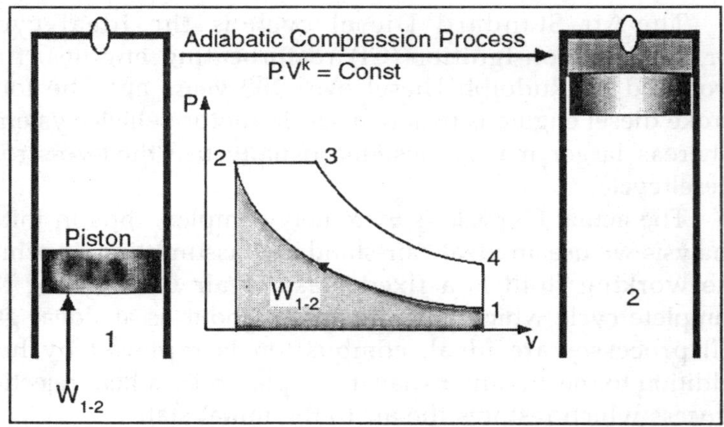

$$w_{1-2} = \int_1^2 P dv = \text{const}$$

$$\int_1^2 V^{-k} dv = \text{Const}\left(\frac{V^{1-k}}{1-k}\right)_1^2 = P.V^k\left(\frac{V^{1-k}}{1-k}\right)_1^2$$

thus,

$$W_{1-2} = \left(\frac{P.V}{1-k}\right)_1^2 = \left(\frac{P_2V_2 - P_1V_1}{(1-k)}\right) = \left(\frac{m.R.(T_2 - T_1)}{(1-k)}\right)$$

An alternative approach using the energy equation takes advantage of the adiabatic process ($Q_{1\text{-}2} = 0$) results in a much simpler process: During process 2-3 the fuel is injected and combusted and this is represented by a constant pressure expansion process.

At state 3 ("fuel cutoff") the expansion process continues adiabatically with the temperature decreasing until the expansion is complete. Process 3-4 is thus the adiabatic expansion process.

The total expansion work is $W_{exp} = (W_{2-3} + W_{3-4})$ and is shown as the area under the *P-V* diagram and is analysed as follows:

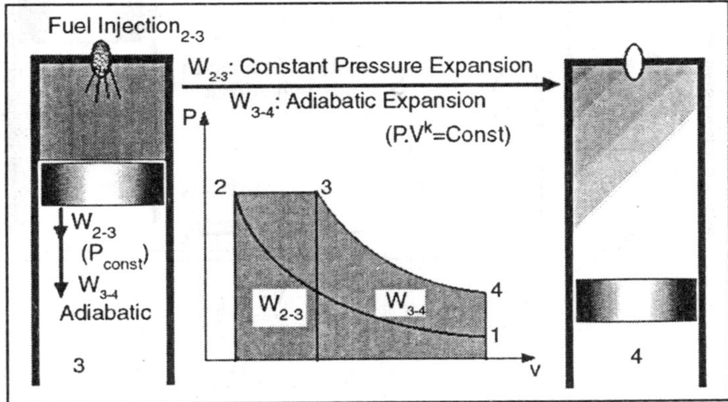

Finally, process 4-1 represents the constant volume heat rejection process. In an actual Diesel engine the gas is simply exhausted from the cylinder and a fresh charge of air is introduced.

The net work W_{net} done over the cycle is given by: $W_{net} = (W_{exp} + W_{1-2})$, where as before the compression work W_{1-2} is negative (work done *on* the system).

In the Air-Standard Diesel cycle engine the heat input Q_{in} occurs by combusting the fuel which is injected in a controlled manner, ideally resulting in a constant pressure expansion process 2-3 as shown below. At maximum volume (bottom dead centre) the burnt gasses are simply exhausted and replaced by a fresh charge of air. This is represented by the equivalent constant volume heat rejection process $Q_{out} = -Q_{4-1}$.

Both processes are analysed as follows:

$$Q_{in} = \Delta H = m \cdot C_p \cdot \Delta T = m \cdot C_p \cdot (T_3 - T_2)$$

$$P \cdot V = m \cdot R \cdot T \Rightarrow \frac{T_3}{T_2} = \frac{V_3}{V_2} \text{ (Constant Pressure)}$$

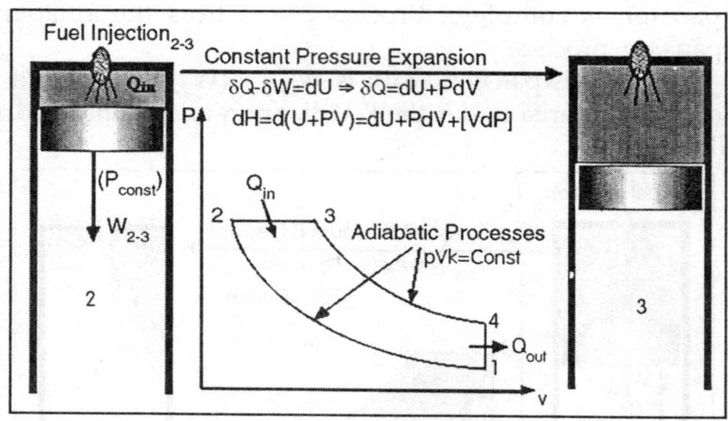

$$Q_{in} = m \cdot C_p \cdot T_2 \cdot \left(\frac{V_3}{V_2} - 1\right)$$

$$= m \cdot C_p \cdot T_2 \cdot (r_c - 1)$$

$$r_c = \frac{V_3}{V_2} \text{ (cut off ratio)}$$

Constant volume heat rejection Q out:

$$Q_{out} = -Q_{4-1} = -\Delta U = -m \cdot C_v \cdot \Delta T = m \cdot C_v \cdot (T_4 - T_1)$$

At this stage we can conveniently determine the engine efficiency in terms of the heat flow as follows:

$$Q_{in} = m \cdot C_p \cdot (T_3 - T_2) \qquad \text{(constant Pressure)}$$

$$Q_{out} = m \cdot C_v \cdot (T_4 - T_1) \qquad \text{(constant volume)}$$

Again from the first law for a cycle:

$$W_{net} = W_{1-2} + W_{2-3} + W_{3-4} = Q_{in} - Q_{out}$$

Thus thermal efficiency,

$$\eta th = \frac{W_{net}}{Q_{in}} = \left(1 - \frac{Q_{out}}{Q_{in}}\right) \longrightarrow$$

Problem

A frictionless piston-cylinder device contains 0.2 kg of air at 100 kPa and 27°C. The air is now compressed slowly according to the relation $P\ V\text{k}$ = constant, where k = 1.4, until it reaches a final temperature of 77°C.

- Sketch the *P-V* diagram of the process with respect to the relevant constant temperature lines, and indicate the work done on this diagram.
- Using the basic definition of boundary work done determine the boundary work done during the process [-7.18 kJ].
- Using the energy equation determine the heat transferred during the process [0 kJ], and verify that the process is in fact adiabatic.

Derive all equations used starting with the basic energy equation for a non-flow system, the equation for internal energy change for an ideal gas (), the basic equation for boundary work done, and the ideal gas equation of state [$P.V = m.R.T$]. Use values of specific heat capacity defined at 300K for the entire process.

Problem

Consider the expansion stroke only of a typical Air Standard Diesel cycle engine which has a compression ratio of 20 and a cutoff ratio of 2. At the beginning of the process (fuel injection) the initial temperature is 627°C, and the air expands at a constant pressure of 6.2 MPa until cutoff (volume ratio 2:1). Subsequently the air expands adiabatically (no heat transfer) until it reaches the maximum volume.

- Sketch this process on a *P-v* diagram showing clearly all three states. Indicate on the diagram the total work done during the entire expansion process.
- Determine the temperatures reached at the end of the constant pressure (fuel injection) process [1800K], as well as at the end of the expansion process [830K], and draw the three relevant constant temperature lines on the *P-v* diagram.

- Determine the total work done during the expansion stroke [1086 kJ/kg].
- Determine the total heat supplied to the air during the expansion stroke [1028 kJ/kg].

Derive all equations used starting from the ideal gas equation of state and adiabatic process relations, the basic energy equation for a closed system, the internal energy and enthalpy change relations for an ideal gas, and the basic definition of boundary work done by a system (if required).

Use the specific heat values defined at 1000K for the entire expansion process, obtained from the table of Specific Heat Capacities of Air.

Solved Problem

An ideal air-standard Diesel cycle engine has a compression ratio of 18 and a cutoff ratio of 2. At the beginning of the compression process, the working fluid is at 100 kPa, 27°C (300 K).

Determine the temperature and pressure of the air at the end of each process, the net work output per cycle [kJ/kg], and the thermal efficiency.

Note

The nominal specific heat capacity values used for air at 300K are,

C_P = 1.00 kJ/kg.K,

C_v = 0.717 kJ/kg.K,,

and k = 1.4.

However they are all functions of temperature, and with the extremely high temperature range experienced in Diesel engines one can obtain significant errors. In this example we develop a temperature averaging technique using the temperature dependence table ofSpecific Heat Capacities of Air.

SOLUTION APPROACH

The first step is to draw a diagram representing the problem, including all the relevant information. We notice that neither volume nor mass is given, hence the diagram and solution will be in terms of specific quantities.

The most useful diagram for a heat engine is the *P-v*diagram of the complete cycle:

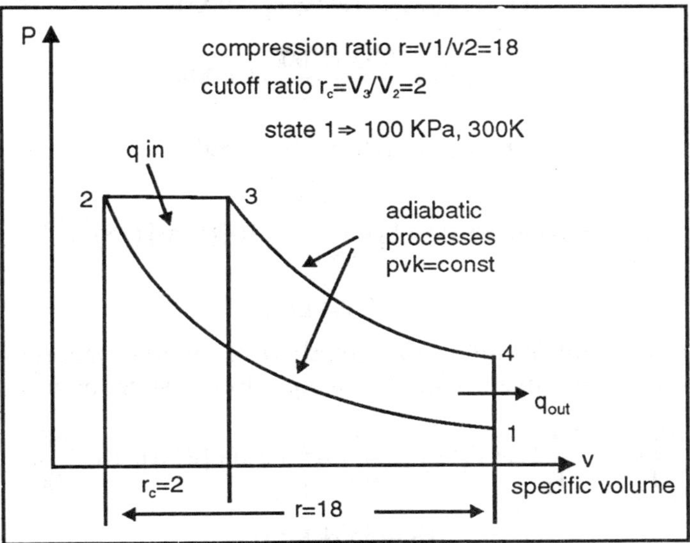

The next step is to define the working fluid and decide on the basic equations or tables to use. In this case the working fluid is air, and we have decided that we will be using temperature average values of specific heat capacities for each process presented in the table of Specific Heat Capacities of Air.

$$Pv = RT, \quad R = 0.287\ [KJ/Kg\ K]$$

$$\Delta u = C_v \Delta T$$

$$\Delta h = C_p \Delta T$$

$$\text{adiabatic} \Rightarrow Pv^k = \text{Const}$$

$$\Rightarrow Tv^{k-1} = \text{Const}$$

We now go through all four processes in order to determine the temperature and pressure at the end of each process.

$$1 \to 2 - \text{Adiabatic compression}$$

$$Tv^{k-1} = \text{const} \Rightarrow T_2 = T_1\left(\frac{v_1}{v_2}\right)^{k-1} = T_1 r^{k-1}$$

$$\text{Try} : K_{300K} = 1.4 \Rightarrow T_2 = 953K$$

$$\text{Try} : \text{Kat}\left(\frac{953+300}{2}\right) \approx 600k$$

$$K_{600k} = 1.376 \Rightarrow T_2 = 889K$$

$$Pv^K = \text{const} \Rightarrow P_2 = P_1\left(\frac{v_1}{v_2}\right)^k = 100[kPa]\,(18)^{1.376}$$

$$P_2 = 5336\,kPa$$

Note that an alternative method of evaluating pressure P_2 is to simply use the ideal gas equation of state, as follows:

$$\frac{P_2 v_2}{T_2} = \frac{P_1 v_1}{T_1} \Rightarrow P_2 = P_1\left(\frac{v_1}{v_2}\cdot\frac{T_2}{T_1}\right) = 100[kPa]\left(18\cdot\frac{889}{300}\right)$$

$$P_2 = 5334\ kPa$$

Either approach is satisfactory - choose whichever you are more comfortable with. We now continue with the fuel injection consant pressure process:

Notice that even though the problem requests "net work output per cycle" we have only calculated the heat in and heat out. In the case of a Diesel engine it is much simpler to evaluate the heat values, and we can easily obtain the net work from the energy balance over a complete cycle, as follows:

$$w_{net} = q_{in} - q_{out} = (1058 - 446) = 612\,kJ/kg$$

Thermal Efficiency:

$$\eta th = \frac{w_{net}}{q_{in}} = \frac{q_{in} - q_{out}}{q_{in}} = 1 - \frac{q_{out}}{q_{in}} = 1 - \frac{446}{1058}$$

$$\eta th = 58\%$$

Chapter 3

Second Law of Thermodynamics

DEFINITION

In this chapter we consider a more abstract approach to heat engine, refrigerator and heat pump cycles, in an attempt to determine if they are feasible, and to obtain the limiting maximum performance available for these cycles. The concept of mechanical and thermal reversibility is central to the analysis, leading to the ideal Carnot cycles.

We represent a heat engine and a heat pump cycle in a minimalist abstract format as in the following diagrams. In both cases there are two temperature reservoirs TH and TL, with TH > TL.

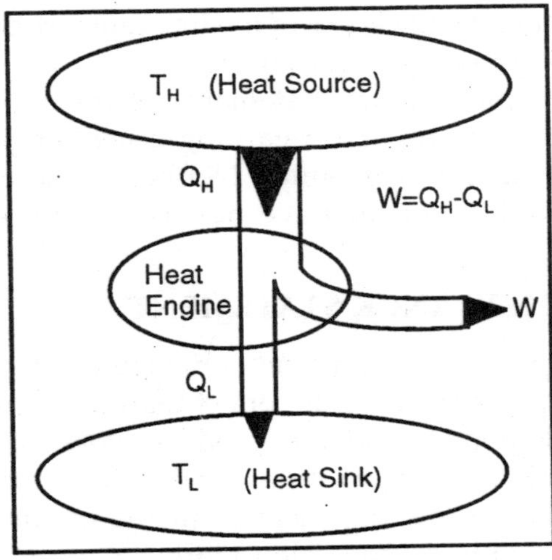

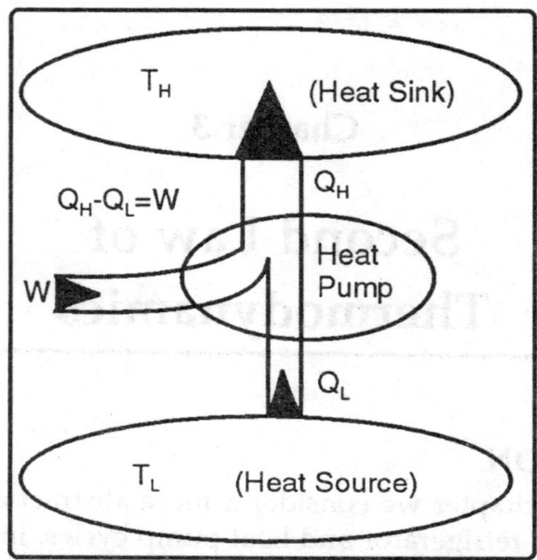

In the case of a heat engine heat QH is extracted from the high temperature source TH, part of that heat is converted to work W done on the surroundings, and the rest is rejected to the low temperature sink TL.

The opposite occurs for a heat pump, in which work W is done on the system in order to extract heat QL from the low temperature source TL and "pump" it to the high temperature sink TH.

Notice that the thickness of the line represents the amount of heat or work energy transferred.

We now present two statements of the Second Law of Thermodynamics, the first regarding a heat engine, and the second regarding a heat pump. Neither of these statements can be proved, however have never been observed to be violated.

THE KELVIN-PLANCK STATEMENT

It is impossible to construct a device which operates on a cycle and produces no other effect than the transfer of heat from a single body in order to produce work.

We prefer a less formal description of this statement in terms of a boat extracting heat from the ocean in order to produce its required propulsion work:

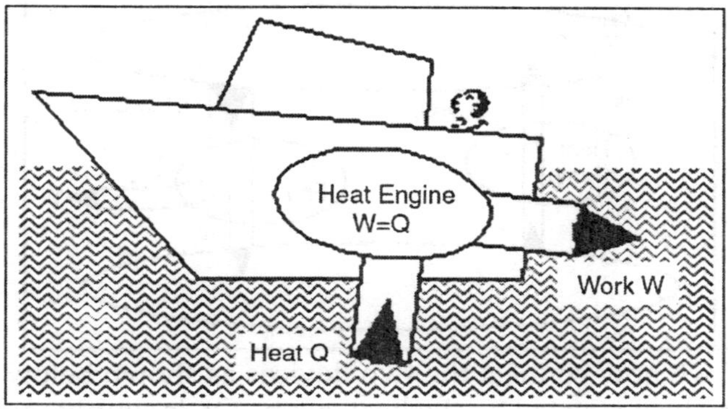

THE CLAUSIUS STATEMENT

It is impossible to construct a device which operates on a cycle and produces no other effect than the transfer of heat from a cooler body to a hotter body.

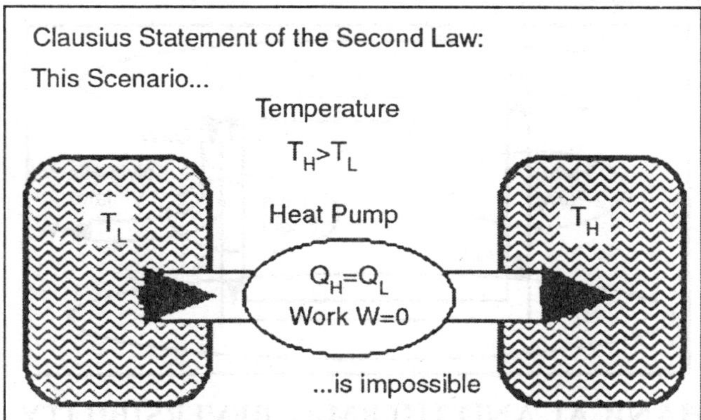

Equivalence of the Clausius and Kelvin-Planck Statements: It is remarkable that the two above statements of the Second Law are in fact equivalent. In order to demonstrate their equivalence consider the following diagram.

On the above we see a heat pump which violates the Clausius statement by pumping heat QL from the low temperature reservoir to the high temperature reservoir without any work input.

On the right we see a heat engine rejecting heat QL to the low temperature reservoir.

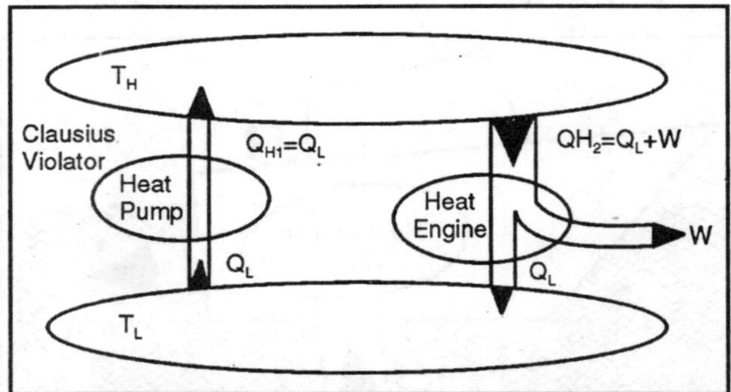

If we now connect the two devices as shown below such that the heat rejected by the heat engine QL is simply pumped back to the high temperature reservoir then there will be no need for a low temperature reservoir, resulting in a heat engine which violates the Kelvin-Planck statement by extracting heat from a single heat source and converting it directly into work.

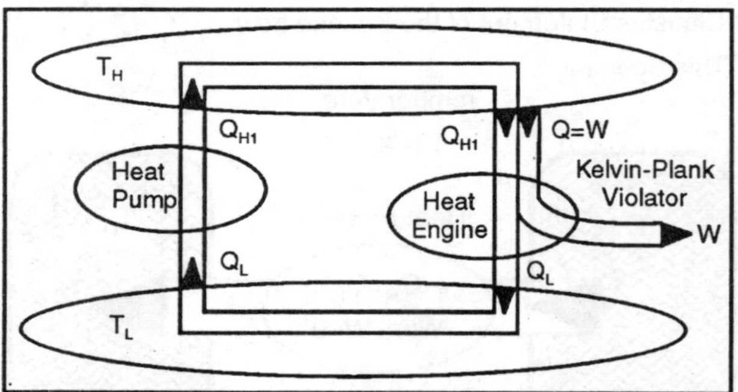

MECHANICAL AND THERMAL REVERSIBILITY

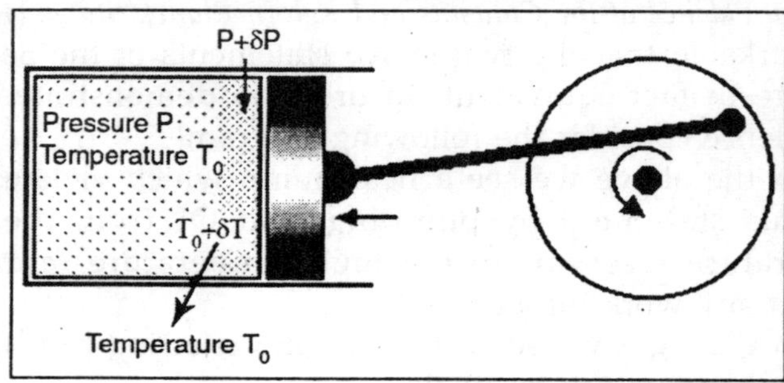

Notice that the statements on the Second Law are negative statements in that they only describe what is impossible to achieve. In order to determine the maximum performance available from a heat engine or a heat pump we need to introduce the concept of Reversibilty, including both mechanical and thermal reversibility. We will attempt to clarify these concepts in terms of the following example of a reversible piston cylinder device in thermal equilibrium with the surroundings at temperature T0, and undergoing a cyclic compression/expansion process.

For mechanical reversibility we assume that the process is frictionless, however we also require that the process is a quasi-equilibrium one. In the diagram we notice that during compression the gas particles closest to the piston will be at a higher pressure than those farther away, thus the piston will be doing more compression work than it would do if we had waited for equilibrium conditions to occur after each incremental step.

Similarly, thermal reversibility requires that all heat transfer is isothermal. Thus if there is an incremental rise in temperature due to compression then we need to wait until thermal equilibrium is established. During expansion the incremental fall in temperature will result in heat being transferred *from* the surroundings *to* the system until equilibrium is established.

In summary, there are three conditions required for reversible operation:

- All mechanical processes are frictionless.
- At each incremental step in the process thermal and pressure equilibrium conditions are established.
- All heat transfer processes are isothermal.

CARNOT'S THEOREM

Carnot's theorem, also known as Carnot's rule, or the Carnot principle, can be stated as follows:

No heat engine operating between two heat reservoirs can be more efficient than a reversible heat engine operating between the same two reservoirs.

The simplest way to prove this theorem is to consider the scenario shown below, in which we have an irreversible engine as well as a reversible engine operating between the reservoirs TH and TL, however the irreversible heat engine has a higher efficiency than the reversible one. They both draw the same amount of heat QH from the high temperature reservoir, however the irreversible engine produces more work WI than that of the reversible engine WR.

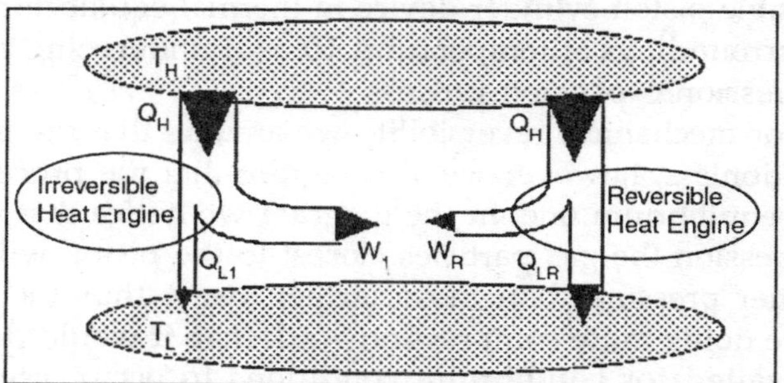

Note that the reversible engine by its nature can operate in reverse, ie if we use some of the work output (WR) from the irreversible engine in order to drive the reversible engine then it will operate as a heat pump, transferring heat QH to the high temperature reservoir, as shown in the following diagram:

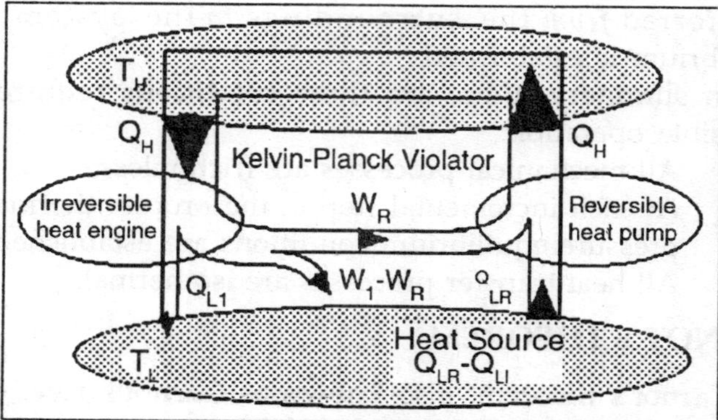

Notice that the high temperature reservoir becomes redundent, and we end up drawing a net amount of heat (QLR - QLI) from the low temperature reservoir in order to produce a net amount of work (WI - WR) - a Kelvin-Planck violator - thus proving Carnot's Theorem.

Corollary 1 of Carnot's Theorem

The first Corollary of Carnot's theorem can be stated as follows: All reversible heat engines operating between the same two heat reservoirs must have the same efficiency.

Thus regardless of the type of heat engine, the working fluid, or any other factor if the heat engine is reversible, then it must have the same maximum efficiency. If this is not the case then we can drive the reversible engine with the lower efficiency as a heat pump and produce a Kelvin-Planck violater as above.

Corollary 2 of Carnot's Theorem

The second Corollary of Carnot's theorem can be stated as follows: The efficiency of a reversible heat engine is a function only of the respective temperatures of the hot and cold reservoirs. It can be evaluated by replacing the ratio of heat transfers QL and QH by the ratio of temperatures TL and TH of the respective heat reservoirs.

Thus using this corollary we can evaluate the thermal efficiency of a reversible heat engine as follows: Notice that we always go into "meditation mode" before replacing the ratio of heats with the ratio of absolute temperatures, which is only valid for reversible machines. The simplest conceptual example of a reversible heat engine is the Carnot cycle engine, as described in the following diagram:

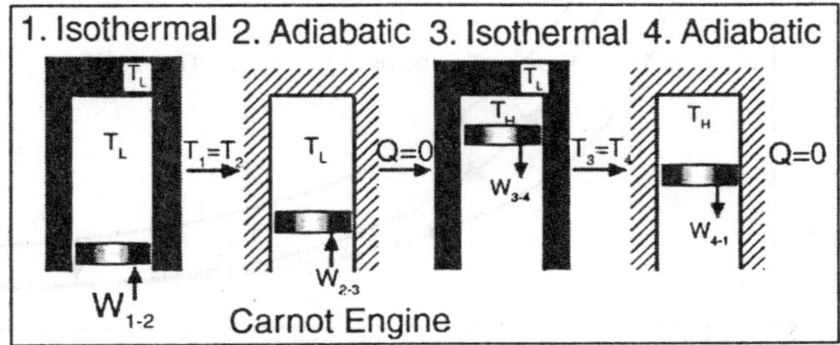

Obviously a totally impractical engine which cannot be realized in practice, since for each of the four processes in the cycle the surrounding environment needs to be changed from isothermal to adiabatic. A more practical example is the ideal Stirling cycle engine as described in the following diagram:

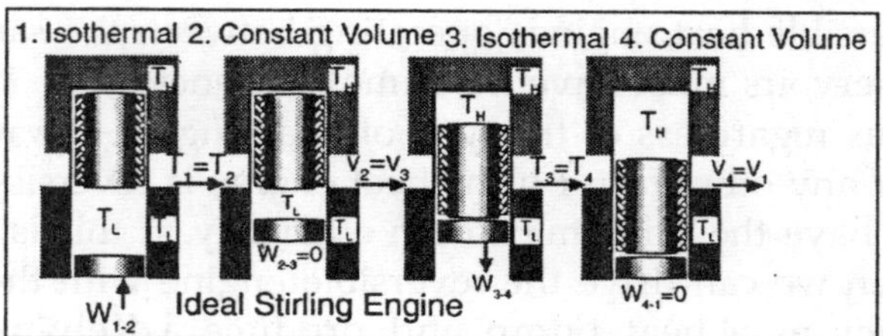

This engine has a piston for compression and expansion work as well as a displacer in order to shuttle the working gas between the hot and cold spaces, and was described previously in . Note that under the same conditions of temperatures and compression ratio the ideal Carnot engine has the same efficiency however a significantly lower net work output per cycle than the Ideal Stirling cycle engine, as can be easily seen in the following diagram:

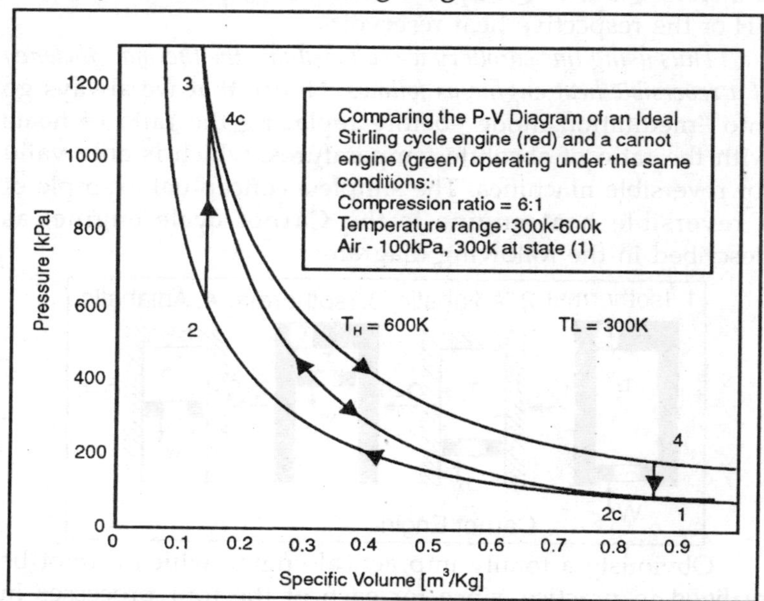

When the reversible engine is operated in reverse it

becomes a heat pump or a refrigerator. Problem a heat pump is used to meet the heating requirements of a house and maintain it at 20°C. On a day when the outdoor air temperature drops to -10°C it is estimated that the house looses heat at the rate of 10 kW. Under these conditions the actual Coefficient of Performance (COPHP) of the heat pump is 2.5.

- Draw a diagram representing the heat pump system showing the flow of energy and the temperatures, and determine:
- The actual power consumed by the heat pump [4 kW]
- The power that would be consumed by a reversible heat pump under these conditions [1.02 kW]
- The power that would be consumed by an electric resistance heater under these conditions [10 kW]
- Comparing the actual heat pump to the reversible heat pump determine if the performance of the actual heat pump is feasible,

Derive all equations used starting from the basic definition of COPHP.

Problem: During an experiment conducted in senior lab at 25°C, a student measured that a Stirling cycle refrigerator that draws 250W of power has removed 1000kJ of heat from the refrigerated space maintained at -30°C. The running time of the refrigerator during the experiment was 20min. Draw a diagram representing the refrigerator system showing the flow of energy and the temperatures, and determine if these measurements are reasonable [COPR = 3.33, COPR,rev = 4.42, ratio COPR/COPR,rev = 75% > 60% - not feasible]. State the reasons for your conclusions. *Derive* all equations used starting from the basic definition of the Coefficient of Performance of a refrigerator (COPR).

LAWS OF HEAT POWER

The Second Law of Thermodynamics is one of three Laws of Thermodynamics. The term "thermodynamics" comes from two root words: "thermo," meaning heat, and "dynamic," meaning power. Thus, the Laws of Thermodynamics are the Laws of "Heat Power." As far as we can tell, these Laws are absolute. All things in the observable universe are affected by and obey the Laws of Thermodynamics.

The First Law of Thermodynamics, commonly known as the Law of Conservation of Matter, states that matter/energy cannot be created nor can it be destroyed. The quantity of matter/energy remains the same. It can change from solid to liquid to gas to plasma and back again, but the total amount of matter/energy in the universe remains constant.

INCREASED ENTROPY

The Second Law of Thermodynamics is commonly known as the Law of Increased Entropy. While quantity remains the same (First Law), the quality of matter/energy deteriorates gradually over time. How so? Usable energy is inevitably used for productivity, growth and repair. In the process, usable energy is converted into unusable energy. Thus, usable energy is irretrievably lost in the form of unusable energy.

"Entropy" is defined as a measure of unusable energy within a closed or isolated system (the universe for example). As usable energy decreases and unusable energy increases, "entropy" increases. Entropy is also a gauge of randomness or chaos within a closed system. As usable energy is irretrievably lost, disorganization, randomness and chaos increase.

IMPLICATIONS OF LAW

The implications of the Second Law of Thermodynamics are considerable. The universe is constantly losing usable energy and never gaining. We logically conclude the universe is not eternal. The universe had a finite beginning — the moment at which it was at "zero entropy" (its most ordered possible state). Like a wind-up clock, the universe is winding down, as if at one point it was fully wound up and has been winding down ever since. The question is who wound up the clock?

The theological implications are obvious. NASA Astronomer Robert Jastrow commented on these implications when he said,

"Theologians generally are delighted with the proof that the universe had a beginning, but astronomers are curiously upset. It turns out that the scientist behaves the way the rest of us do when our beliefs are in conflict with the evidence."

KELVIN-PLANCK AND CLAUSSIUS STATEMENTS

STATEMENT

Kelvin-Planck's statement is based on the fact that the efficiency of the heat engine cycle is never 100%. This means that in the heat engine cycle some heat is always rejected to the low temperature reservoir.

The heat engine cycle always operates between two heat reservoirs and produces work.

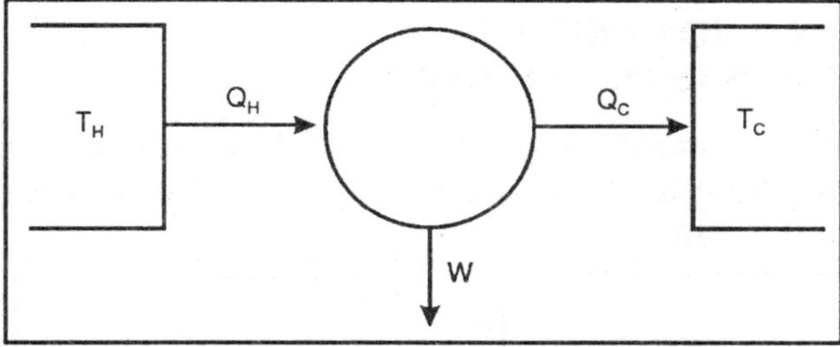

The statement made by Kelvin-Planck for third law of thermodynamics says, "It is impossible for a heat engine to produce net work in a complete cycle if it exchanges heat only with bodies at a single fixed temperature." Thus to produce the work the cycle should exchange heat with two reservoirs which are a different temperatures. The high temperature reservoir is called as source and low temperature reservoir is called as sink.

$$W = Q_H - Q_C$$

As per the above statement the net work will be produced in the cycle as long as there is difference in temperature between the source and sink.

In due course of time if source loses too much heat and sink gains too much heat and their temperatures become equal, the net work produced in the cycle will be zero.

The Kelvin-Planck statement tells the condition for producing work within the cycle, the Carnot's statement tells maximum work or efficiency that can be obtained within the cycle.

The ratio of the maximum mechanical work obtain to the total heat supplied to the engine is known as maximum thermal efficiency (ς_{max}) of the engine.

Mathematically,

ς_{max} = Maximum work obtain/Total heat supplied

$$= Q_H - Q_C/Q_H = 1 - Q_C/Q_H = 1 - T_C/T_H$$

For a reversible engine,

$$Q_H/T_H = Q_C/T_C$$

KELVIN AND CLAUSIUS STATEMENTS OF 2ND LAW

The two statements of the second law may not appear to have anything to do with one another, but in fact each one implies the other.

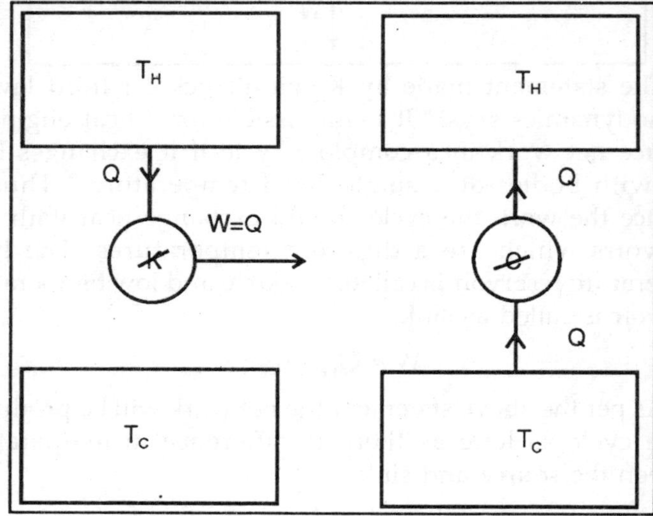

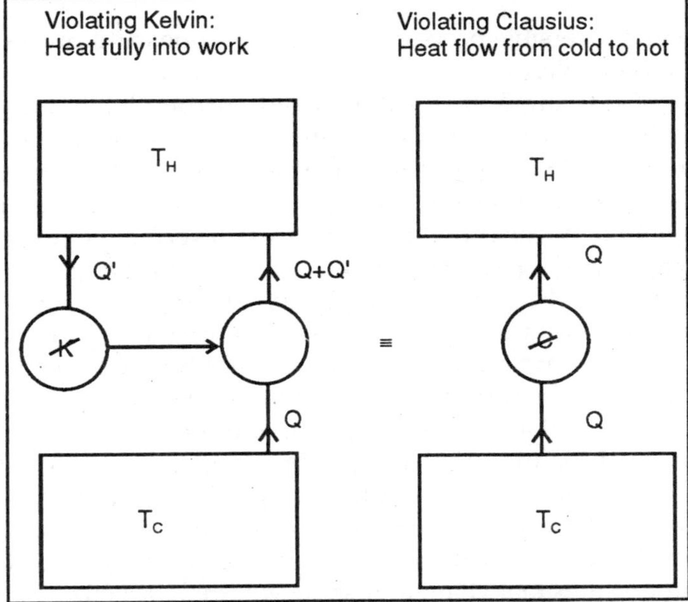

The top line shows yellow engines representing hypothetical engines violating each of the two statements of the second law. The second line shows that an engine which violates Kelvin's statement, together with a normal heat pump, violates Clausius's statement. A similar drawing can be used to show that an pump which violates Clausius's statement, together with a normal heat engine, violates Kelvin's statement. Hence the two statements—which seem quite different—are actually equivalent.

REVERSIBLE PROCESSES

DEFINITION

A reversible process describes an ideal thermodynamic systemwhere steps in a reaction, or changes in a physical process, can reverse direction such that the resulting entropy change for thesystem and surroundings is zero. If it is possible to restore a thermodynamic system and its surroundings to their original state (including the original entropic state) following a process then such a process is termed a reversible process. All other reactions and processes are irreversible.

Considering both thesystem and surroundings, all real processes are irreversible. Internallyreversible processes demand no irreversibilities in the system and externally reversibleprocess demand no irreversibilities outside thesystem. A ideal reversible processdescribes a reaction orprocess in which thesystem its surroundings are in equilibrium throughout the process. In considering ideal gasexpansions and contractions, for example, the system and surroundings are considered to be in equilibrium if they have the same temperature and the same pressure throughout theprocess. If at any time the system or the surrounds diverge from equilibrium then the process becomes irreversible thermodynamically. The definitional (and easily constructed) test of reversibility dictates that if a reaction or process is completelyreversible then it can be driven to change direction by an infinitesimal change in some external variable.

To expand an ideal gas by a reversible process demands that outside and inside pressures be kept balanced as the expansion proceeds. The temperatures of system and surroundings must remain the same so that the system and surroundings are in equilibrium at all times. The only way to accomplish this is through a series of thermostatic steps in which the system does a maximum amount work on the surroundings (expand) while the surroundings does minimum work on the system.

Although the second law of thermodynamics states that all isolated systems move spontaneously toward a state of maximum entropy, as a practical matter, nearly reversible process can be designed if reaction or process conditions are not allowed to stray far from equilibrium. Because entropy increases create irreversible process and set a limiting boundary on the approach to reversibility, reversible processes must be designed to reduce entropy increase to a minimum. Physicists studying the thermodynamics of reversible processes or equilibrium dynamicsuse thermostatic techniques to create near reversible processes through a series of slow and gradual sequence steps conducted at, or very near, the equilibrium point of each step. In some cases, thermostatic techniques substitute a series of very slow and gradual changes into the transition from initial to final states.

In many idealized approximations (especially at the high school and undergraduate university level) students encounter idealized problems that are reversible. Very few, if any, real problems are so idealized and entropic and chaotic effects make natural reactions or processes irreversible.

Most reactions are irreversible because energy is usually transformed and work is done to cause a change in a system or its surroundings. In additon, in most reactions some energy is lost because of undesirable transformations of energy (e.g., heat lost during the use of electrical energy). No matter how well a systemis insulated, or designed to reduce such factors as friction, there is always energy dissipation and entropy increase.

By definition, reversible process demand 100% efficiency in the use and conversion of energy (any loss would preclude a return to the initial state). Because heat (generated by friction, etc.) can never be 100% recovered and reconverted into work, no machine can operate at 100% efficiency. As a consequence, machines can not perforn reversible functions. Imagine a cylinder, with a perfectly smooth piston, which contains gas. If you push with a force only just large enough to overcome the internal pressure, the volume will start to decrease slowly. Then if you decrease the force only slightly, the volume will start to increase. This is the hallmark of a reversible process: an infinitesimal change in the external conditions reverses the direction of the change. Heat flow is only reversible if the temperature difference between the bodies is infinitesimally small.

Reversible processes require the absence of friction or other hysteresis effects. They must also be carried out infinitesimally slowly. Otherwise pressure waves and finite temperature gradients will be set up in the system, and irreversible dissipation and heat flow will occur.

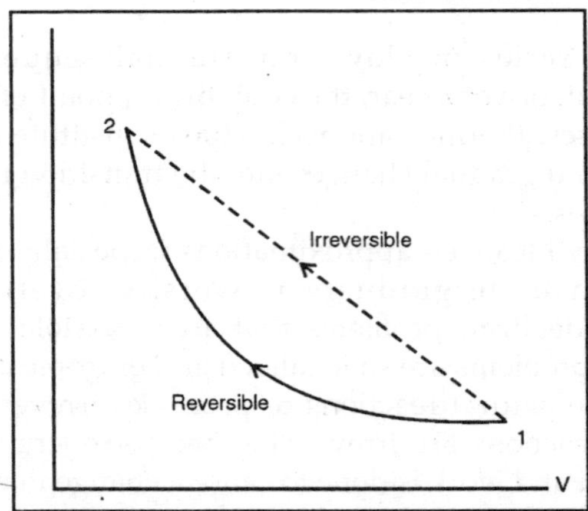

Because reversible processes are very slow, the system is always very nearly in equilibrium at all times. In that case all its state variables are well defined and uniform, and the state of the system can be represented on a plot of, for instance, pressure versus volume. A finite reversible process passes through an infinite set of such states, and so can be drawn as a solid line on the plot. During an irreversible process the system is not in an equilibrium state, and so cannot be represented on the plot; an irreversible process is often drawn as a straight dotted line joining the initial and final equilibrium states. The work done during the process is *not* then equal to the area under the line, but will be greater than that for the corresponding reversible process.

REVERSIBLE AND IRREVERSIBLE PROCESS

When the system undergoes change from its initial state to the final state, the system is said to have undergone a process. During thermodynamic process the one or more of the properties of the system like temperature, pressure, volume, enthalpy or heat, entropy etc changes. The second law of thermodynamics enables to classify all the processes under two main categories: reversible or ideal process and irreversible or natural process.

REVERSIBLE PROCESS

The process in which the system and surroundings can

be restored to the initial state from the final state without producing any changes in the thermodynamics properties of the universe is called as the reversible process. In figure below, let us suppose that the system has undergone change from state A to state B. If the system can be restored from state B to state A, and there is no change in the universe, then the process is said to be reversible process. The reversible process can be reversed completely and there is no trace left to show that the system had undergone thermodynamic change.

For the system to undergo reversible change, it should occur infinitely slowly due to infinitesimal gradient. During reversible process all the changes in state occurred in the system are in thermodynamic equilibrium with each other. Thus there are two important conditions for the reversible process to occur. Firstly, the process should occur in infinitesimally small time and secondly all the initial and final state of the system should be in equilibrium with each other.

If during the reversible process the heat content of the system remains constant i.e. it is adiabatic process, then the process is also isentropic process i.e. the entropy of the system remains constant. The phenomenon of undergoing reversible change is also called as reversibility. In actual practice the reversible process never occurs, thus it is an ideal or hypothetical process.

Process

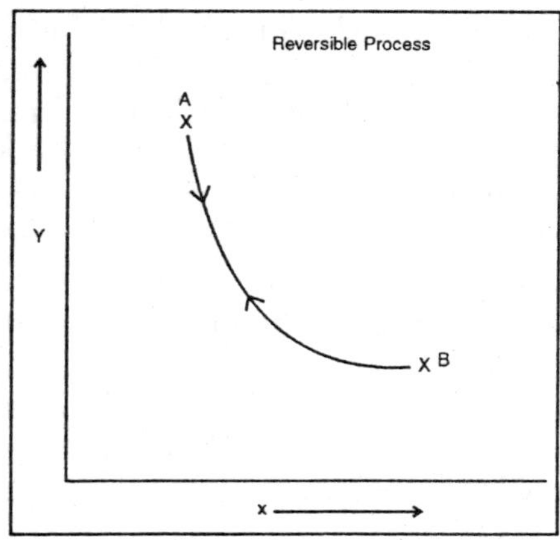

IRREVERSIBLE PROCESS

The irreversible process is also called as natural process because all the processes occurring in the nature are irreversible processes. The natural process occurs due to finite gradient between the two states of the system. For instance heat flow between two bodies occurs due to temperature gradient between the two bodies; this is in fact the natural flow of heat.

Similarly, water flows from high level to low level, current moves from high potential to low potential etc.

Here are some important points about the irreversible process:

- In the irreversible process the initial state of the system and surroundings cannot be restored from the final state.
- During irreversible process the various states of the system on the path of change from initial state to final state are not in equilibrium with each other.
- During the reversible process the entropy of the system increases decisively and it cannot be reduced back to its initial value.
- The phenomenon of system undergoing irreversible process is called as irreversibility.

FEATURES OF REVERSIBLE PROCESSES

Reversible processes are idealizations or models of real processes. One familiar and widely used example is Bernoulli's equation, which you saw in Unified. They are extremely useful for defining limits to system or device behaviour, for enabling identification of areas in which inefficiencies occur, and in giving targets for design.

An important feature of a reversible process is that, depending on the process, it represents the maximum work that can be extracted in going from one state to another, or the minimum work that is needed to create the state change.

Let us consider processes that do work, so that we can show that the reversible one produces the maximum work of all possible processes between two states. For example, suppose we have a thermally insulated cylinder that holds an ideal gas, Figure. The gas is contained by a thermally insulated massless piston with a stack of many small weights on top of it. Initially the system is in mechanical and thermal equilibrium.

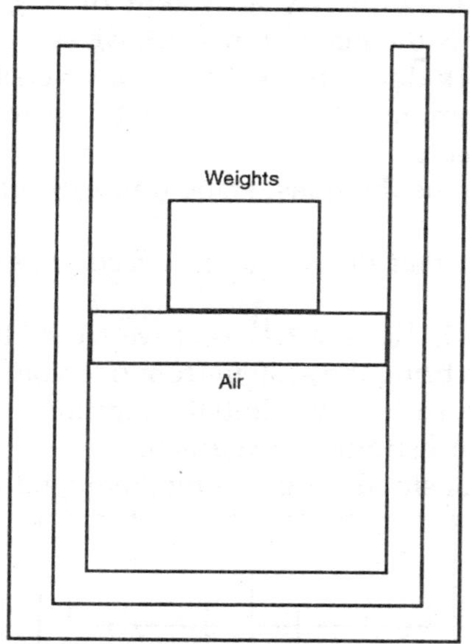

Fig. A Piston with Weights on Top

Consider the following three processes, shown in Figure:

- All of the weights are removed from the piston instantaneously and the gas expands until its volume is increased by a factor of four (a free expansion).
- Half of the weight is removed from the piston instantaneously, the system is allowed to double in volume, and then the remaining half of the weight is instantaneously removed from the piston and the gas is allowed to expand until its volume is again doubled.
- Each small weight is removed from the piston one at a time, so that the pressure inside the cylinder is always in equilibrium with the weight on top of the piston. When the last weight is removed, the volume has increased by a factor of four.

- Maximum work (proportional to the area under these curves) is obtained for the quasi-static expansion.

To Reiterate:

- The work done by a system during a reversible process is the maximum work we can get.
- The work done on a system in a reversible process is the minimum work we need to do to achieve that state change.

A process must be quasi-static (quasi-equilibrium) to be reversible.

This means that the following effects must be absent or negligible:

- *Friction*: If $P_{external} \neq P_{system}$ we would have to do net work to bring the system from one volume to another and return it to the initial condition.
- Free (unrestrained) expansion.
- Heat transfer through a finite temperature difference.

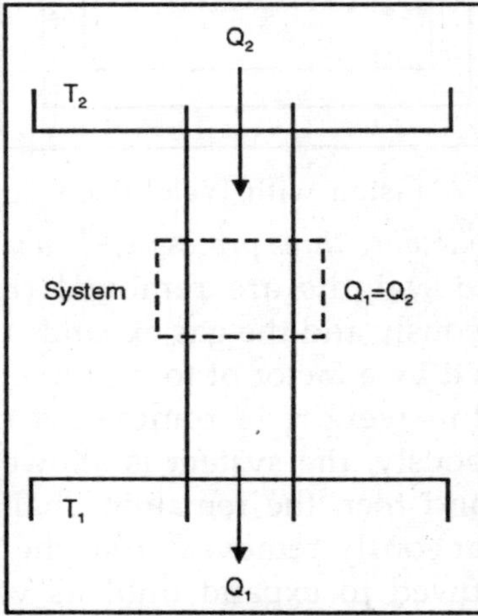

Fig. Heat Transfer Across a Finite Temperature Difference

Suppose we have heat transfer from a high temperature to a lower temperature as shown in Fig. How do we restore the situation to the initial conditions? One thought would be to run a Carnot refrigerator to get an amount of heat, Q from the lower temperature reservoir to the higher temperature reservoir. We could do this but the surroundings, again us, would need to provide some amount of work (which we could find using our analysis of the Carnot refrigerator). The net (and only) result at the end of the combined process would be a conversion of an amount of work into heat. For reversible heat transfer from a heat reservoir to a system, the temperatures of the system and the reservoir must be,

$$T_{heat\,reservoir} = T_{system} \pm dT \ .$$

In other words the difference between the temperatures of the two entities involved in the heat transfer process can only differ by an infinitesimal amount, dT.

While all natural processes are irreversible to some extent, it cannot be emphasized too strongly that there are a number of engineering situations where the effect of irreversibility can be neglected and the reversible process furnishes an excellent approximation to reality. The second law, which is the next topic we address, allows us to make a quantitative statement concerning the irreversibility of a given physical process.

CARNOT CYCLES

Take-home message: The efficiency of a Carnot engine is independent of its construction, and no irreversible engine can beat it. A Carnot engine is, simply, a reversible engine acting between only two heat reservoirs. That means that all processes are either isothermal (heat transfer at a constant temperature) or adiabatic (no heat transfer). By contrast the Otto cycle, which has heating at constant volume, would need a whole series of heat reservoirs at incrementally higher temperatures to carry out the heating reversibly.

Carnot's theorem says that a reversible engine is the most efficient engine which can operate between two reservoirs. If you want to see the proof, An equally important corollory is that any reversible engine working between two heat reservoirs has the same efficiency as any other, irrespective of the details of the engine.

Much is made of the fact that the Carnot engine is the most efficient engine. Actually, this is not mysterious. First, if we specify only two reservoirs, then all it says is that a reversible engine is more efficient than an irreversible engine, which isn't surprising (no friction...) Second, we will see that the efficiency of a Carnot engine increases with the temperature difference between the reservoirs. So it makes sense to use only the hottest and coldest heat baths you have available, rather than a whole series of them at intermediate temperatures. But the independence of the details of the engine is rather deeper, and has far-reaching consequences.

As a result, if we can calculate the efficiency for one Carnot engine, we know it for all. We can calculate it for an ideal gas Carnot cycle, and find,

$$\eta_{carnot} = 1 - \frac{T_C}{T_H}$$

Hence this is true for *all* Carnot engines.

By comparing with the definition of the efficiency of any heat engine, $\eta = W / Q_H = 1 - Q_C / Q_H$, we get the even more useful relation:

$$\frac{Q_C}{Q_H} = \frac{T_C}{T_H}$$

Carnot developed the concept of reversibility and showed that no engine could be more efficient than a reversible one before either the first or second law of thermodynamics had been formulated! .

CARNOT WINS

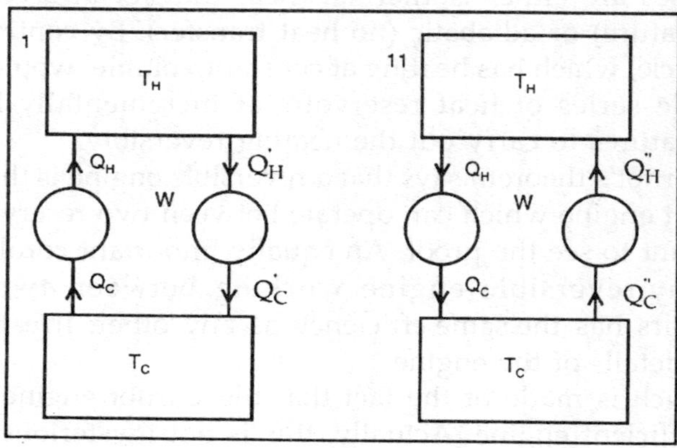

In the above the green engines/pumps are Carnot, and the brown ones are irreversible.

Because the Carnot engine is reversible, if the engine efficiency is η_C, the pump efficiency is $1/\eta_C$.

- In the first case, a Carnot pump is driven by an engine. Overall, to avoid violating Clausius's statement of the second law, there must be no net heat transfer to the hot reservoir, so

$$Q_H' \geq Q_H .$$

Also, by conservation of energy,

$$Q_H - Q_C = Q_H' - Q_C' = W .$$

Thus,

$$\frac{W}{Q_H'} = \frac{Q_H}{Q_H'}\frac{W}{Q_H}$$

$$\text{so } \eta_{engine} \leq \eta_{carnot}$$

- In the second case, a Carnot engine drives a pump. Now we need

$$Q_H \geq Q_H'' .$$

Thus,

$$\frac{Q_H''}{W} = \frac{Q_H''}{Q_H}\frac{Q_H}{W}$$

$$\text{so } \eta_{pump} \leq \frac{1}{\eta_{carnot}}$$

So no engine can be more efficient than a Carnot engine, and no pump can be more efficient than a Carnot pump. If the brown pumps were in fact reversible, there could be no overall heat flow, since heat flow from a hot to a cold body is an irreversible process. In that case the inequalities would become equalities. Thus any reversible engine working between two heat reservoirs has the same efficiency as any other.

EFFICIENCY OF IDEAL GAS CARNOT CYCLE

Above we see the four steps which make up an ideal gas Carnot cycle, two isothermal, (i) and (iii), and two adiabatic, (ii) and (iv). Below the cycle is sketched on a P-V plot.

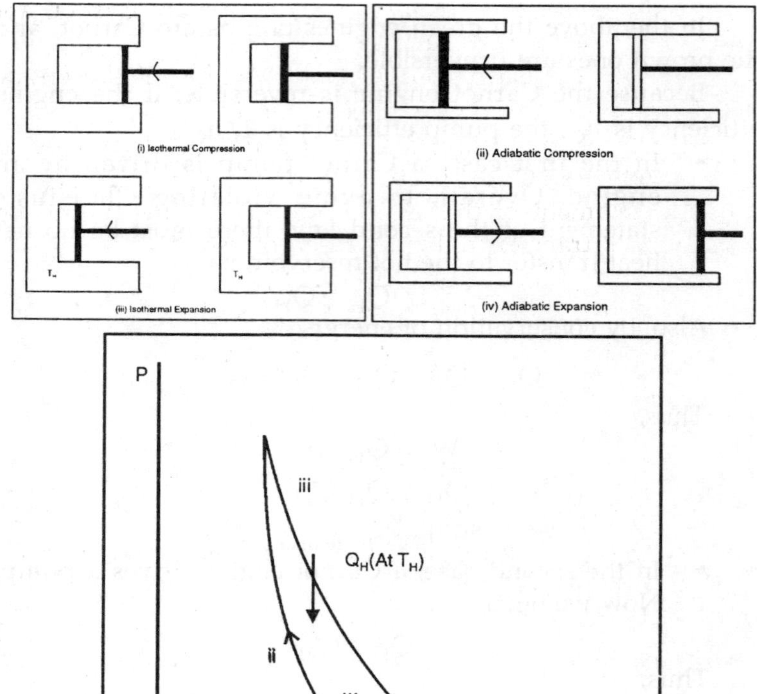

We have already analysed this particular cycle We worked out the work for the four steps. We note further that the heat exchanged during the isothermal steps is just minus the work done, since the internal energy is unchanged.

Thus, we have

$W = nR \ln 2(T_H - T_0)$ and $Q_H = nRT_H \ln 2$, so

$$\eta = \frac{W}{Q_H} = 1 - \frac{T_0}{T_H}$$

Note this cycle wasn't general, because we chose to make both the isothermal and adiabatic stages volume-halving. But form of the answer is general.

CARNOT'S THEOREM

DEFINITION

Lazare Nicolas Marguérite Carnot was a person of many gifts and strong political views. During the Revolution he directed the Army of the North and under Napoleon Bonaparte he served as the Minister of War. In 1794, together with Gaspard Monge he founded the now famous Ecole Polytechnique. His geometric and military publications are considered masterpieces in their respective areas.

The nice little theorem that bears his name shows an unexpected relationship between the radii of the circumcircle and incircle of a given triangle.

In the formulation and the proof of Carnot's theorem I'll use the standard notations. The theorem read as follows:

In any triangle ABC, the (algebraic) sum of the distances (suitably signed) from the circumcentre O to the sides, is R + r, the sum of the circumradius and the inradius.

$$OM_a + OM_b + OM_c = R + r.$$

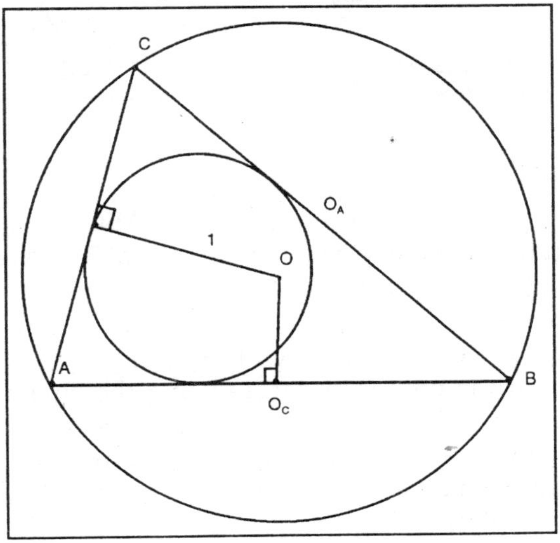

In acute triangles, the circumcentre is always located inside the triangle. In this case, all three segments OM_a, OM_b, and OM_c lie entirely inside the triangle.

If one of the angles is obtuse, the circumcentre falls outside the triangle. One of the segments (the one that corresponds to the side opposite the obtuse angle) lies entirely outside, while the other two segments lie only partially outside the triangle.

In the above sum, the segments that intersect the interior of the triangle are taken with the sign *plus*, the remaining side with the sign *minus*. Segments OM_a, OM_b, and OM_c serve as altitudes of triangles OBC, OAC, and OAB with the bases on the sides a, b, and c. The sign convention guarantees that the areas of these triangles (taken with the proper signs) always add up to the area of ΔABC.

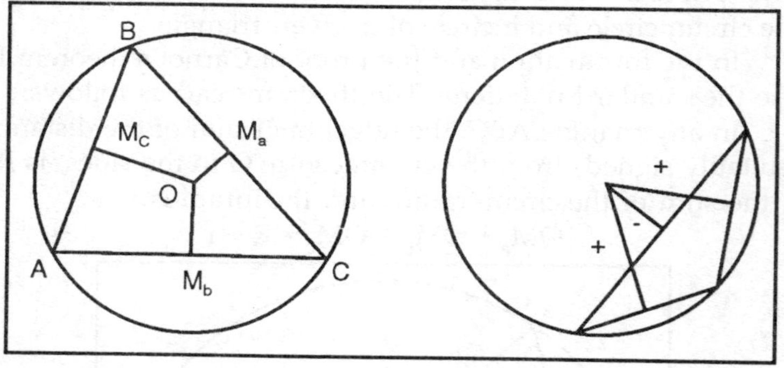

PROOF

For the inradius we always have,

$$r(a + b + c) = 2Area(ABC)$$

From the above, we also have Area(OAB) + Area(OBC) + Area(OAC) = Area(ABC).

Therefore,

$$(*) r(a + b + c) = aOM_a + bOM_b + cOM_c$$

A remark is now in order. O being the circumcentre of ABC, in the isosceles triangle AOB, AOB = 2C. And similarly, BOC = 2A and AOC = 2B. Equipped with this knowledge we may consider several triples of similar (right-angled) triangles:

- ABH_b, ACH_c, and BOM_a (or, an equal COM_a)
- BAH_a, BCH_c, and COM_b (or, an equal AOM_b)
- CBH_b, CAH_a, and AOM_c (or, an equal BOM_c)

From the first triple we derive,

$AH_b/c = AH_c/b = OM_a/R$

which leads to,

$OM_a(b + c) = R(AH_b + AH_c)$

Similarly, we get two additional identities,

$OM_b(a + c) = R(BH_a + BH_c)$

$OM_c(a + b) = R(CH_a + CH_b)$

Summing the three up yields after simplifications,

$OM_a(b + c) + OM_b(a + c) + OM_c(a + b) = R(a + b + c)$

Adding this to (*) and dividing by (a + b + c) completes the proof.

Carnot's theorem is a direct application of the theorem of Pythagoras.

It's a nice little theorem that is unlikely to be rediscovered with dynamic geometry software. A penchant for doodling and imagination this is all that is probably needed to rediscover a theorem like that.

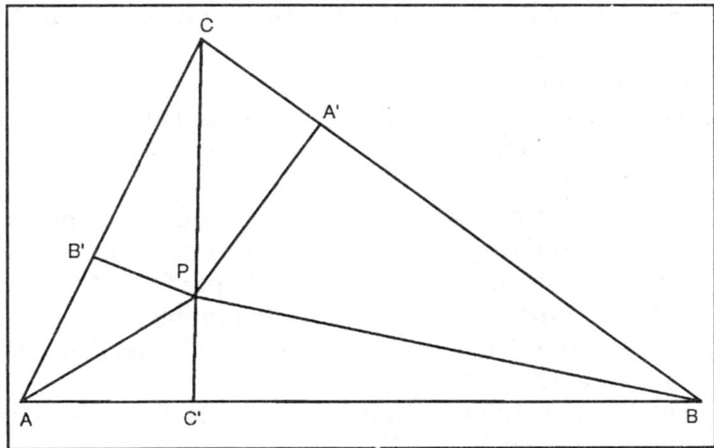

Let points A′, B′, and C′ be located on the sides BC, AC, and respectively AB of ΔABC. The perpendiculars to the sides of the triangle at points A′, B′, and C′ are concurrent if,

$$AC'^2 - BC'^2 + BA'^2 - CA'^2 + CB'^2 - AB'^2 = 0.$$

PROOF

Assume the three perpendiculars meet in point P. The lines AP, BP, CP, A'P, B'P, C'P split the triangle into six right-angled triangles, three pairs of which share a hypotenuse, whilst three other pairs share a leg with a vertex at P.

The situation begs for an application of the Pythagorean theorem:

$AC'^2 + C'P^2 = AP^2$

$BC'^2 + C'P^2 = BP^2$

$BA'^2 + A'P^2 = BP^2$

$CA'^2 + A'P^2 = CP^2$

$CB'^2 + B'P^2 = CP^2$

$AB'^2 + B'P^2 = AP^2$.

Taking the first, third, and fifth equations with the sign "+", and the second, fourth, and sixth equations with the sign "-", and adding up all six leads, after obvious simplifications. To proof the converse, assume (1) and let P be the point of intersection of, say, A'P and B'P. Drop a perpendicular DP from P to AB.

By the already proven part:

(1)$AD^2 - BD^2 + BA'^2 - CA'^2 + CB'^2 - AB^2 = 0$,

so that

(2) $AD^2 - BD^2 = AC'^2 - BC'^2$.

This is only possible if D = C'. Why so. Let BD = x, AB = c. Then AD = c - x, and

$$AD^2 - BD^2 = (c - x)^2 - x^2 = c^2 - 2cx,$$

which is a linear function of x: $f(x) = c^2 - 2cx$. f therefore is 1-1. In other words, f(x) can't have the same value for two different x's. This means that (2) does imply D = C'.

We are naturally coming to a related problem: Given two points A and B. What is the locus of points M such that $AM^2 - BM^2$ is constant?

Let M be a point on the locus and H the foot of the perpendicular from M to AB.

By the Pythagorean theorem:

$AM^2 - AH^2 = MH^2 = BM^2 - BH^2$,

i.e.,

$AM^2 - BM^2 = AH^2 - BH^2$,

which means that, if M lies on the locus in question, so does H. And vice versa.

Since the same H serves as the foot of the perpendicular MH for all points M of the line perpendicular to AB at H, this line solves the problem. If we place the origin of a Cartesian coordinates system at B with the x-axis along AB and the y-axis perpendicular to it, we may assign coordinates (x, y) to point M.

Then,

$BM^2 = x^2 + y^2$, whereas

$AM^2 = (c - x)^2 + y^2$.

Thus, as above,

$AM^2 - BM^2 = (c - x)^2 + y^2 - x^2 - y^2 = f(x)$,

where f(x) is the function defined above. If x is fixed, so is f(x), and also $AM^2 - BM^2$, as expected. From this fact, or from the underlying proof, it is clear that, for (1) to hold, the points A′, B′ C′ need not be necessarily located on the sides of ABC.

COROLLARY

1. Perpendicular bisectors of the sides of a triangle concur in a point (the circumcentre.)
   Indeed, AC′ = BC′, BA′ = CA′, and CB′ = AB′ imply (1).
2. The altitudes of a triangle concur in a point.
   Indeed, if AA′, BB′, and CC′ are altitudes of ABC, then
   $AB^2 - BA'^2 = AA'^2 = AC^2 - CA'^2$,
   so that,
   $AB^2 - AC^2 = BA'^2 - CA'^2$,
   Similarly,
   $AC^2 - BC^2 = AC'^2 - BC'^2$ and
   $BC^2 - AB^2 = CB'^2 - AB'^2$, which,
   when added, give (1).
3. The perpendiculars to the sides of ABC at the points where the incircle meets the sides are concurrent (at the incentre.)
4. The perpendiculars to the sides of ABC at the points

where the excircles meet the sides are concurrent. Indeed [F. G.-M., p. 555], in this case we have
CA′ = AC′, BA′ = AB′, and BC′ = CB′,
which implies (1).

5. Existence of Orthopole

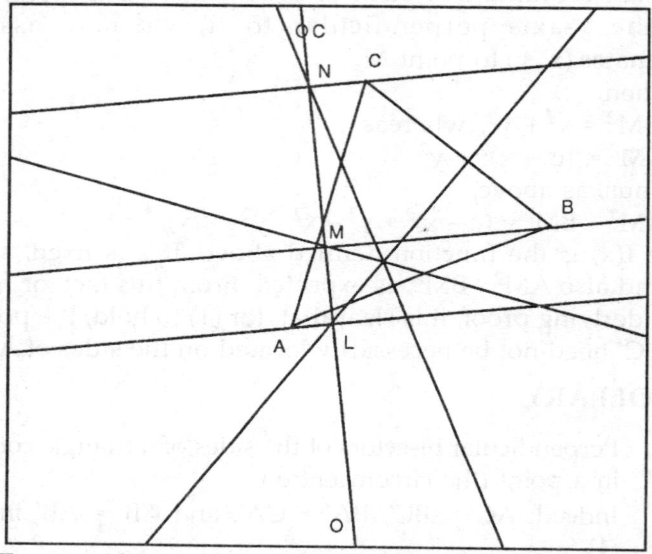

To every line m in the plane of ΔABC there correspond a point - the orthopole of M with respect to ΔABC. To define the point, first drop perpendiculars AL, BM and CN from the vertices of ΔABC onto m. From the three points thus obtained drop perpendiculars on the "opposite" sides of the triangle: from L onto BC, from M onto AC, and from N onto AB. The latter three lines are concurrent, and the point where they meet is known as the *orthopole*of m with repsect to ΔABC.

Darij Grinberg came up with an ingenious proof of this result based on Carnot's theorem (1). As we already know, the points A′, B′, C′ need not be located on the sides of ΔABC. So let's take A′ = L, B′ = M, and C′ = N. Applying the Pythagorean theorem several times, we obtain,

$AM^2 - LM^2 = AL^2 = AN^2 - LN^2$,

so that,

$AM^2 - AN^2 = LM^2 - LN^2$.

Similarly,

$BN^2 - BL^2 = MN^2 - LM^2$, and

$CL^2 - CM^2 = LN^2 - MN^2$,

which, when added, produce

$AM^2 - AN^2 + BN^2 - BL^2 + CL^2 - CM^2 = 0.$

ENTROPY

CONCEPT

The concept of entropy is fundamental to understanding the second law of thermodynamics. Entropy (or more specifically, increase in entropy) is defined as heat (in calories or Btu's) absorbed by a system, divided by the absolute temperature of the system at the time the heat is absorbed. Absolute temperature is the number of degrees above "absolute zero", the coldest temperature that can exist.

The total entropy in a system is represented by the symbol ΔS. The symbol S is used to represent a given change in the entropy content of a system. If the symbol q is used to represent the amount of heat absorbed by a system, the equation for the resulting entropy increase is:

$$\Delta S = q/T \quad (1)$$

Where T is the absolute temperature. When heat is absorbed, the entropy of a system increases; when heat flows out of a system, its entropy decreases.

The "surroundings" of a system is everything outside of the system that can interact with it; surroundings can usually be defined as the space that surrounds a system. When heat is evolved by a system, that same heat is absorbed by its surroundings.

When heat is absorbed by a system, that same heat must necessarily come from its surroundings. Therefore any entropy increase in a system due to heat flow must be accompanied by an entropy decrease in the surroundings, and vice versa. When heat flows spontaneously from a hotter region to a cooler region, the entropy decrease in the hotter region will always be less than the entropy increase in the cooler region, because the greater the absolute temperature, the smaller the entropy change for any particular heat flow.

As an example, consider the entropy change when a large rock at 500 degrees absolute is dropped into water at 650 degrees absolute. (We are using an absolute temperature scale based on Fahrenheit degrees; on this scale, water freezes at 492 degrees.) For each Btu of heat that flows into the rock at these temperatures the entropy increase in the rock is 1/500 = 0.0020 and the entropy decrease of the water is 1/650 = 0.0015. The difference between these values is 0.0020 - 0.0015 = 0.0005. This represents the *over all* entropy increase of the system (rock) and its surroundings (water).

Of course the rock will warm up to, and the water cool to, a temperature intermediate between their original temperatures, thus considerably complicating the calculation of total entropy change after equilibrium is achieved. Nevertheless, for every Btu of heat transferred from water to rock there will always be an increase of over-all net entropy.

As was mentioned before, a spontaneous change is an irreversible change. Therefore an increase in the overall net entropy can be used as a measure of the irreversibility of spontaneous heat flow.

Irreversible changes in a system can, and often do, take place even though there may be no interaction, and negligible heat flow, between system and surroundings. In cases like these the entropy "content" of the system is greater after the change than before. Even when heat flow does not occur between system and surroundings, spontaneous changes inside an isolated system are always accompanied by an increase in the system's entropy, and this calculated entropy increase can be used as a measure of irreversibility. The following paragraphs will explain how this entropy increase can, at least in some cases, be calculated.

It is an axiom of thermodynamics that entropy, like temperature, pressure, density, etc., is a property of a system and depends only on the existing condition of the system. Regardless of the procedures followed in achieving a given condition, the entropy content for that condition is always the same. In other words, for any given set of values for pressure, temperature, density, composition, etc., there can be only one value for the entropy content. It is essential to remember this: When a system that has undergone an irreversible change is restored to its original condition (same temperature, pressure, volume, etc.) its entropy content will likewise be the same as it was before.

In cases where an isolated system undergoes an entropy increase as the result of a spontaneous change inside the system, we can calculate that entropy increase by postulating a procedure whereby the system's entropy increase is transferred to the surroundings in a manner such that there is no further increase in *net* entropy and the system is restored to its original condition. The entropy increase of the surroundings can then be readily calculated by equation:

(1) DS = q/T, where q = heat absorbed by the surroundings, and T = absolute temperature of the surroundings.

It bears repeating that when the system is restored to its original condition, its entropy content will be the same as it was before its irreversible change. Therefore the amount of entropy absorbed by the surroundings during restoration must necessarily be the same as the entropy increase accompanying the system's original irreversible change, providing that there is no further increase in *net* entropy during restoration.

This postulated restoration procedure and the postulated properties of the surroundings are for the purpose of calculation only. Since we are not dealing with the surroundings as such, they can be postulated in whatever form necessary to simplify the calculations; it is neither necessary nor desirable that the surroundings correspond to any condition that could actually exist. Therefore, we will postulate a theoretical restoration procedure that takes place with no further increase in *net* entropy, even though such a procedure can not actually be obtained experimentally.

The restoration process, if it were to take place in actuality, would have to be accompanied by at least a small amount of irreversibility, and hence an additional increase in the entropy of the surroundings beyond the entropy increase from the system's original irreversible change. This is because heat will not flow without a temperature differential, friction cannot be entirely eliminated, etc. Therefore the restoring process, if it is to take place with no further increase in overall net entropy, must be postulated to take place with no irreversibility.

If such a process could be actually realized, it would be characterized by a continuous state of equilibrium (i.e. no pressure or temperature differentials) and would occur at a rate so slow as to require infinite time. Processes like these are called "reversible" processes. Remember, reversible processes are postulated to simplify the calculation of the entropy change in a system; it is *not* necessary that they be capable of being achieved experimentally.

It should not be assumed that equation requires that q, the heat absorbed, must necessarily be absorbed reversibly. The concept of reversibility is merely a means to an end: the calculation of entropy change accompanying an irreversible process.

The following example will illustrate the calculation of a reversible restoring process and at the same time develop the equation which is the basis for the thermodynamical relationship between probability and the second law. We will postulate a system consisting of an "ideal" gas contained in a tank connected to a second tank that has been completely evacuated, with the valve between the two tanks closed. The temperature of the system and its surroundings is postulated to be the same.

An ideal gas is one whose molecules are infinitely small and have no attractive or repulsive forces on each other. (Under ordinary conditions hydrogen and helium closely approximate the properties of an ideal gas.) An ideal gas is chosen in order to develop the basic relationship without introducing complicating correction factors to account for the size of the molecules and the forces they exert on each other.

When the valve is opened the gas expands irreversibly from V1 (its original volume) to V2 (the volume of both tanks).

There is no work of compression by or upon the surroundings. Because the gas is ideal there is no temperature change, and hence no heat flow takes place.

After expanding irreversibly from V1 to V2, the gas is restored to its original condition by reversibly compressing it back to V1. This compression requires work (force applied through a distance) which in turn generates heat in the gas, heat that is absorbed by the surroundings so that there is no increase in the gas temperature. In our mathematical model of this reversible restoring process, the surroundings are postulated to be so large that they also do not undergo any temperature increase. The temperature T remains unchanged during the entire irreversible expansion and subsequent reversible restoration process. The work of compressing the gas during restoration is equal to the pressure of the gas times the volume change due to compression. Because the pressure increases during compression, the work of compression must be determined by the calculus integral:

$$\text{Compression Work} = \int PdV$$

where: P = pressure

V = volume

dV = the small change in volume taking place at the corresponding pressure P

The integral sign indicates the summation of all the individual values of PdV.

The equation relating temperature, pressure, and volume of an ideal gas is:

$$PV = RT \ (2)$$

where: P = pressure

V = volume

T = absolute temperature

R = a constant which depends only on the amount of gas present

In the case of a reversible, isothermal compression of an ideal gas we may substitute P from equation (2) into the equation for compression work. When this is done, *we have*:

$$\text{Compression Work} = RT\int dV/V \quad (3)$$

Although it is not necessary that our postulated reversible restoration process be capable of being carried out in a practical sense, it is nevertheless sometimes helpful to be able to visualize the process. To this end, the reader may consider the restoring compression process being brought about by a piston fitted into the end of the second tank. On compression from V2 to V1, the piston moves down the length of the second tank, and with no mechanical friction forces all the gas contained therein back into the first tank V1.

Since the work of compression is equal to q, the heat absorbed by the surroundings, q may be substituted in equation (3) to give:

$$q = RT\int dV/V \quad (4)$$

From equation (1) the entropy gained by the surroundings during restoration from V2 to V1 is:

$$\Delta S = q/T \quad (1)$$

Substituting from equation (4):

$$\Delta S = RdV/V$$

Upon integrating (a calculus procedure for summing up the individual values of dV/V) we have:

$$\Delta S = R\ln(V2/V1) \quad (5)$$

Where ln(V2/V1) is the natural logarithm of the ratio of expanded volume to the initial volume, and ΔS is equal to the entropy increase in the surroundings upon restoration compression from V2 to V1. As we have seen, ΔS is also equal to the entropy increase of the gas caused by its original expansion from V1 to V2. This is because V1 is the same volume both before expansion and after restoration compression, and therefore has the same entropy content. Therefore the entropy transferred to the surroundings during restoration is equal to that gained by the system in expanding from V1 to V2.

ENTROPY AND PROBABILITY

The ratio of the probability that all the gas molecules are evenly distributed between the two tanks to the probability that all the molecules, of their own accord and by random motion, would be in tank V1 is equal to $(V2/V1)^N$, where N is the number of molecules.

If V2/V1 were equal to 2.0, for example, and N were equal to 10, the probability ratio would be 2 to the tenth power, or 1024. For N = 100, the ratio would be approximately 1.27 times ten to the 30th power. It is clear that the random motion of trillions of gas molecules heavily favours a uniform distribution.

let X1 = the probability of all the gas molecules, after the valve is opened, remaining in the first tank V1

let X2 = the probability of all the gas molecules, after the valve is opened, being uniformly distributed in V2, the volume of both tanks.

From the probability equation, we have:

$X2/X1 = (V2/V1)^N$.

Taking the natural logarithm of both sides, and then multiplying both sides by R, the gas constant:

$R \ln(X2/X1) = RN \ln(V2/V1)$

$R/N \ln(X2/X1) = R \ln(V2/V1)$

Substituting in equation (5):

$\Delta S = R/N \ln(X2/X1)$ (6)

Equation (6) represents the fundamental relationship between probability and the second law of thermodynamics. It states that the entropy of a gaseous system increases when its molecular distribution changes from a lower probability to a higher probability (X2 greater than X1).

Based on the belief that the laws of thermodynamics are universal, this equation has been assumed to apply to all systems, not just gaseous. In other words, any entropy change is proportional to the logarithm of the ratio of probabilities. Therefore, for the general case equation (6) can be written:

$\Delta S = K \ln(X2/X1)$ (7)

Where K is a constant depending on the particular change involved. However, individual values of K, X1, or X2 are seldom, if ever, known for non-gaseous systems.

As we have seen before, ΔS can be either positive or negative.

When ΔS is negative equation (7) can be written:

$-\Delta S = -K \ln(X2/X1)$

$= K \ln(X1/X2)$

Therefore *a system can go from a more probable state (X2) to a less probable state (X1), providing S for the system is negative.* In cases where the system interacts with its surroundings, S can be negative *providing* the over-all entropy of the system and its interacting surroundings is positive; the over-all change can be positive if the entropy increase of the surroundings is numerically greater than the entropy decrease of the system. In the case of the formation of the complex molecules characteristic of living organisms, creationists raise the point that when living things decay after death, the process of decay takes place with an increase in entropy.

They also point out, correctly, that a spontaneous change in a system takes place with a high degree of probability. They fail to realize, however, that probability is relative, and a spontaneous change in a system can be reversed, providing the system interacts with its surroundings in such a manner that the entropy increase in the surroundings is more than enough to reverse the system's original entropy increase.

The application of energy can reverse a spontaneous, thermodynamically "irreversible" reaction. Leaves will spontaneously burn (combine with oxygen) to form water and carbon dioxide. The sun's energy, through the process of photosynthesis, will produce leaves from water vapour and carbon dioxide, and form oxygen.

If we unplug a refrigerator, heat will flow to the interior from the surroundings; the entropy increase inside the refrigerator will be greater than the entropy decrease in the surroundings, and the net entropy change is positive. If we plug it in, this spontaneous "irreversible" change is reversed. Due to the input of electrical energy to the compressor, the heat transferred to the surroundings from the condenser coils is greater than the heat extracted from the refrigerator, and the entropy increase of the surroundings is greater than the entropy decrease of the interior, in spite of the fact that the surroundings are at a higher temperature. Here again, the net entropy change is positive, as would be expected for any spontaneous process.

In a similar manner, the application of electrical energy can reverse the spontaneous reaction of hydrogen and oxygen to form water: when a current is passed through a water solution, hydrogen is liberated at one electrode, oxygen at the other.

As can easily be confirmed experimentally, agitating water raises its temperature. When water falls freely from a higher elevation to a lower elevation, its energy is changed from potential to kinetic, and finally to heat as it splashes at the end of its fall. The second law of thermodynamics states that the water will not spontaneously raise itself to its original elevation using the heat produced on splashing as the sole source of energy. To do so would require a heat engine that would convert all of the heat of splashing to mechanical energy.

The efficiency of a heat engine is thermodynamically limited by the Carnot cycle, which limits the efficiency of any heat engine to $\Delta T/T$, where ΔT is the temperature increase due to splashing, and T is the absolute temperature. Since ΔT is only a small fraction of T, there is no device that could be constructed which would allow all of the water to spontaneously jump back to its former elevation.

We can, at least in theory, calculate the entropy increase of the water resulting from its irreversible change in falling. In a manner analogous to that used in the previous example, the entropy increase would be equal to the heat generated by splashing agitation, divided by the absolute temperature. If some of the energy of the falling water is extracted by a water wheel, there will be less heat of splashing and hence less entropy increase.

A properly designed turbine could extract most of the water's kinetic energy. This is *not* the same thing as trying to utilize the heat of splashing as an energy source for a heat engine to raise the water. In other words, using the energy before it becomes heat is much more efficient than trying to use it after it becomes heat. If a water wheel is connected by shafts, belts, pulleys, etc. to a pump, the pump can raise water from the downstream side of the water wheel to an elevation even higher than that of the upstream reservoir. *Some* of the water would spontaneously raise itself to an elevation even higher than original, but the rest of it would end up below the water wheel on the downstream side. While it is not possible for *all* of the water to raise itself to an elevation higher than its initial elevation, it is possible for *some* of the water to spontaneously raise itself to an elevation higher than initial.

As with any other irreversible change, there will be an increase in over-all entropy. This means that the entropy increase of the water going over the wheel is greater than the entropy decrease of water pumped up to the higher elevation.

This will be confirmed mathematically in the following paragraphs. will stand for the Greek letter gamma, representing unit weight in pounds per cubic foot. An increase in the value of a parametre will be represented by Δ.

Let:

γ = unit weight of water, lbs./cubic foot

Δh = height of reservoir above downstream side, feet

γh = potential energy of water in reservoir

Δh = additional height above reservoir (to which water is pumped)

$h + \Delta h$ = height to which water is pumped, feet above downstream side

w = work of pumping water to higher elevation

q = heat loss due to pump friction and downstream agitation

f = fraction of water pumped to elevation h + h

From the flow equation, energy in = energy out:

$\gamma h = q + w$

Where:

h = the total energy expended by the water in falling from height h.

q = the energy wasted as heat of splashing

w = work expended in pumping the water to a higher

elevation

$q = T\Delta S$ (from equation 1)

$w = f\gamma(h + \Delta h)$

The total energy available, γh, is divided into pump work, $f\gamma(h + \Delta h)$, and energy lost, $T\Delta S$:

$\gamma h = T\Delta S + f\gamma(h + \Delta h)$

Rearranging:

$T\Delta S = \gamma h - f\gamma(h + \Delta h)$ (8)

When no pump work is done, then:

$T\Delta S' = \gamma h$ (9)

Combining equations (8) and (9), we get:

$\Delta S' - \Delta S = f\gamma(h + \Delta h)/T$ (10)

In the case where the water falls freely without turning the water wheel or operating the pump:

let q'= heat of splashing

$q' = T\Delta S'$

where $\Delta S'$ is the entropy increase on free fall

Equation (10) shows that $\Delta S'$ is larger than ΔS, and that the entropy increase due to pump friction and downstream agitation is "backed up" by the even larger entropy increase that takes place when water falls freely. Equation (10) also shows that the lower the value of ΔS, the more efficient the pump, and the greater the value of f, the fraction of water pumped.

Creationists assume that a change characterized by a decrease in entropy can not occur under any circumstances. In fact, spontaneous entropy decreases can, and do, occur all the time, providing sufficient energy is available. The fact that the water wheel and pump are man-built contraptions has no bearing on the case: thermodynamics does not concern itself with the detailed description of a system; it deals only with the relationship between initial and final states of a given system (in this case, the water wheel and pump).

A favourite argument of creationists is that the probability of evolution occurring is about the same as the probability that a tornado blowing through a junkyard could form an airplane. They base this argument on their belief that changes in living things have a very low probability and could not occur without "intelligent design" which overcomes the laws of thermodynamics. This represents a fundamental contradiction in which (they say) evolution is inconsistent with thermodynamics because thermodynamics doesn't permit order to spontaneously arise from disorder, *but* creationism (in the guise of intelligent design) doesn't have to be consistent with the laws of thermodynamics.

A simpler analogy to the airplane/junkyard scenario would be the stacking of three blocks neatly on top of each other. To do this, intelligent design is required, but stacking does not violate the laws of thermodynamics. The same relations hold for this activity as for any other activity involving thermodynamical energy changes. It is true that the blocks will not stack themselves, but as far as *thermodynamics* is concerned, all that is required is the energy to pick them up and place them one on top of the other.

Thermodynamics merely correlates the energy relationships in going from state A to state B. If the energy relationships permit, the change may occur. If they don't permit it, the change can not occur.

A ball will not spontaneously leap up from the floor, but if it is dropped, it will spontaneously bounce up from the floor. Whether the ball is lifted by intelligent design or just happens to fall makes no difference. On the other hand, thermodynamics does not rule out the possibility of intelligent design; it is just simply not a factor with respect to the calculation of thermodynamic probability. Considering the earth as a system, any change that is accompanied by an entropy decrease (and hence going back from higher probability to lower probability) is possible as long as sufficient energy is available. The ultimate source of most of that energy, is of course, the sun.

The numerical calculation of entropy changes accompanying physical and chemical changes are very well understood and are the basis of the mathematical determination of free energy, emf characteristics of voltaic cells, equilibrium constants, refrigeration cycles, steam turbine operating parametres, and a host of other parametres. The creationist position would *necessarily discard the entire mathematical framework* of thermodynamics and would provide no basis for the engineering design of turbines, refrigeration units, industrial pumps, etc. It would do away with the well-developed mathematical relationships of physical chemistry, including the effect of temperature and pressure on equilibrium constants and phase changes.

PHYSICAL CONCEPT OF ENTROPY

CONSERVATION OF ENERGY

The conservation of energy is the bedrock of our understanding of nature. Conservation of energy, however, only tells us that in any process the total energy remains constant, that is . Energy conservation, however, does not specify what can enter and leave the energy conservation equation, only that the sum of energy entering and leaving a system must always be the same. Furthermore, energy conservation does not tell us what are the various forms of energy, and how energy transforms from one form into another. From our study of waves, light, electric and magnetic fields and so on, we found that energy plays a central role in all these phenomena.

However, only the specific study of the various phenomena could unravel the specific forms that energy takes. The concept of energy also does not tell us what fraction of the total energy is kinetic and what fraction is electrical, chemical and so on. A new concept,in addition to energy, is needed for a more complete understanding of physical processes, in particular those involving the *transformation of energy*. In all the examples we studied, energy could be completely transformed from one form into another.

For example, in our study of magnetic fields, we found that we could fully convert mechanical energy into electrical currents by Faraday's Law of Induction. We may well ask: can all forms of energy be fully transformed from one form to another? The answer is unexpectedly complicated. In some cases the answer is "yes", and in other cases the answer is "no". The concept of entropyaddresses this question regarding the transformation of the forms of energy, and tells us how much energy can be converted from one form into another, and in particular, how much energy is available for doing useful work. Heat had defied all attempts to explain it using the concept of force, and, in general using Newtonian mechanics.

The fact that heat is a form of energy presented a great breakthrough in the understanding of heat, and was fundamental to the emergence of energy as one of the most fundamental ideas in physics. It was, however, soon realized that heat was not an ordinary form of energy such as gravitational or electrical energy.

Rather, heat is a unique form of energy in that only a certain fraction of heat energy can be transformed into other forms of energy. It is for understanding the unique problems presented by heat that new concepts such as entropy, which go beyond energy, were developed. The idea of entropy was introduced into physics in the nineteenth century and is essential for understanding the phenomenon of heat. Entropy is derived from the Greek word meaning "transformation content".

Famous names such as Sadi Carnot, R.J. Clausius, Count Rumford, James Joule, Ludwig Boltzmann and so on contributed greatly to the ideas of heat and entropy. Entropy is a concept which does not answer all the questions left unanswered by energy, but it does address a crucial aspect of energy, namely how much energy can be transformed from one form into another. In particular, if a certain amount of heat energy is given, then entropy tells us how much "useful" work can be extracted from heat energy, in other words, what is the maximum possible efficiency of a heat engine.

In general, since energy is conserved, one would think there is no need to be concerned about recklessly wasting energy. But we know from daily life that "useful" energy, or equivalently, energy "available" for use, is a scarce resource. This intuitive understanding that useful energy is a precious resource is explained by the concept of entropy. Entropy is also related to the concept or order and disorder, and a phase transition from say water to gas is permissible only if entropy increases in such a transformation.

For any system, entropy is a physically measurable quantity. Furthermore, it is an experimentally observed fact that for all processes in nature, the entropy of the total system can never decrease. Note this empirical law concerning entropy is weaker than that of energy, since it does not specify, for any physical process, by how much does entropy increase.

The concept of entropy has many other applications and vast ramifications in other disciplines. Only those chemical processes and reactions for which the total entropy of the system does not decrease are allowed by nature. All living entities must maintain their low entropic state to stay alive. More abstract applications of entropy occur in the physics of black holes, and it was only in 1997 that the entropy of black holes could be calculated from first principles using results from string theory.

The concept of entropy also has a key application in information theory. It was shown by Claude Shannon in 1950 that the information content of a message is determined by a suitable application of the idea of entropy to the discipline of information science. The current understanding of entropy is that it is a large scale (macroscopic) manifestation of the atomic nature of matter.

For example, one cubic centimetre of air contains about number of atoms. This is an unimaginably large number of particles.

Even if under some circumstances we can view these atoms as classical particles, it is futile to apply Newton's laws to such a large collection of particles. Instead, the best we can do is to acknowledge our total ignorance and assume that all the atoms are moving randomly, that is, the position and velocity of every atom is a random variable.

Furthermore, for a system of atoms that reaches a stable state, called equilibrium (more on this concept later),every possible velocity and position of the atoms in one cubic centimetre of air is equally likely. \We will show later that the empirical law that entropy never decreases is simply a statement that for a system consisting of a collection of atoms which is not in equilibrium, the collection of atoms will always move towards a state that is the more probable, and will only reach equilibrium on arriving at the most probable state available to the system.

It should be noted that the field of thermodynamics developed in the eighteenth and nineteenth centuries without the concept of the atomic composition of matter.

It is remarkable that very general results on how bulk matter behaves were obtained without any idea of the microscopic composition of matter.

From the point of view of scientific methodology, the progress of thermodynamics shows that there are concepts describing the large scale properties of matter that have a logic of their own, and which later on are seen to match on smoothly to a deeper and more complete understanding of the same phenomenon.

Temperature and Heat

We discuss thermodynamics, which is the branch of physics dealing with heat, because this will give us a specific physical system in which we can precisely define and discuss entropy and related ideas. In particular, we will study the behaviour of of an ideal gas to illustrate the concept of heat, energy and entropy. Most of us have an intuitive understanding of heat, based on having experienced hot and cold objects. The first thing one must realize is that heat is different from temperature.

TEMPERATURE

What do we mean by saying that an object is at temperature ? Temperature is a physical property of all physical systems. From the microscopic point of view, the temperature of a physical object is a measure of the average kinetic energy of the atoms that make up that object. If the atoms, on the average, are moving very fast, they have high kinetic energy and hence the object has a high temperature, and vice versa for slowly moving atoms. We can immediately see that since there is no limit on how much momentum a particle can have, there is consequently no upper limit on how high the temperature can be.

The concept of a maximum temperature does, however, appear in Big Bang cosmology, but it is so large as to be practically infinite. On the other hand, there must be a lower limit to temperature, which is reached when all of the atoms that constitute an object are brought to the state of lowest possible energy, called the ground state. It should be noted that at absolute zero temperature, unlike classical systems is which there is no longer any motion, a quantum system in at absolute zero temperature is in its ground state, and continues to have quantum fluctuations.

Zeroth Law of Thermodynamics

Two objects are in thermal equilibrium if they are at the same temperature. The zeroth law is the basis of the measurement of temperature, and of the existence of thermometres.

By bringing an object of known temperature into contact with the object whose temperature is being measured, once thermal equilibrium is reached one can conclude that both the temperatures are equal.

Any physical material whose behaviour under changes of temperature is known can be used to measure temperature. For example, if we know how a metal expands on being heated, or how the resistance of a conductor changes, can be used for measuring temperature. There is an absolute scale of temperature for which the lowest temperature is at a temperature called absolute zero. This scale is called Kelvin and denoted by . Let be the temperature of the object in the Kelvin scale, and T_C be its temperature on the ordinary Celsius scale. We then have the following.

$$T_C = T - 273.15K \ T_C = T - 273.15K$$

Hence, for example, 20°C=293.15K. Temperature is a new dimensional quantity in addition to M,L and T. The units of temperature is K and this is also used to signify its dimension. The fact that it is much more difficult to cool than to heat an object is evidenced by the fact that refrigerators were only made in the twentieth century whereas heat furnaces and engines have been in existence for many centuries. The lowest temperature reached to date was achieved in 1992 and is

$$T = 2 \times 10^{-9} K$$

For every extra decimal point of cooling closer to absolute zero, new experimental techniques are required and are usually based on qualitatively new physical properties on various phenomena. A table of temperatures found in nature is given in Table

Table Some Typical Temperatures of Physical Phenomena

Characteristic Temperatures		K
Universe at the Big Bang	10^{30}	
Highest Laboratory Temperature	10^{9}	
Surface of Sun		5×10^{3}
Normal Human Temperature	310	
Freezing Temperature of Water	273	
Universe today	3	
Boiling of Helium -3		10^{-1}
Lowest Laboratory Temperature	10^{-9}	

HEAT

Every object, at temperature T, has an internal, thermal, energy denoted by E, and which can be thought of as being composed of the sum of the kinetic and potential energy of all the atoms which constitute the object. Heat, denoted by Q, is a form of energy, called thermal energy since it is a form arising from temperature. Heat always refers to a process in which thermal energy is exchanged between objects at different temperatures. Heat is analogous to work in that both heat and work are defined by a process, be it the process of the flow of heat, or the process of a force acting over some distance. Furthermore, neither heat nor work are intrinsic properties of the system and cannot be stored in the system;

what can be stored instead is energy, and which is reflected in the total energy of the system. The dimension of heat is that of energy, and the unit of heat is calorie, defined to be the energy required to heat 1 gm of water by 1°C. Since J is also a unit of energy, we have the following conversion.

1 Calorie = 4.186J

If two objects having different temperatures are brought into contact such that heat flows between them - called thermal contact - then heat will flow from the object at the higher temperature to that which is at lower temperature, until both the objects reach a common temperature.

The final state is said to be anequilibrium state in which there are no more macroscopic changes taking place. A very important concept in thermodynamics is that of the heat bath, which is taken to be a large reservoir maintained at some constant temperature .

An example of a heat bath is a swimming pool maintained at some constant temperature. When an object is brought into thermal contact with the heat bath, it either gains heat if it is cooler, or loses heat if it is hotter, than ; the important point to note is that in the heat flows, the heat bath is taken to be so big that its temperature does not change, regardless of its exchange of heat with the object in question.

Heat flow is taken to be positive if energy flows into the object and adds to the internal energy E, and is taken to be negative if a part of the internal energy flows out of the system.

Let the system's temperature be . Hence, if $T_S > T_E$, the heat Q that flows is positive, and if $T_S > T_E$, then the heat flow is negative. Figure shows this flow of heat.

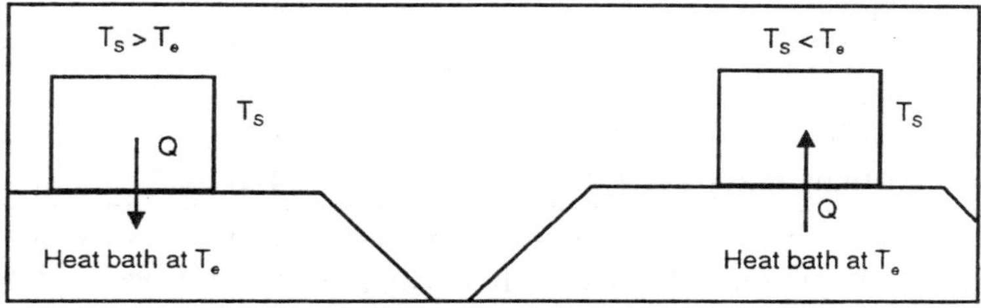

Fig. Heat Flows of Object and Environment

The specific heat of an object, say c, determines how much the temperature of an object, of mass m, changes if it absorbs Q amount of heat. If the initial temperature of an object is T_i and after absorbing Q amount of heat its final temperature is T_f, then the definition of c is given by,

$$T_f - T_i = \frac{Q}{mc}$$

$$\Rightarrow c = \frac{Q}{m(T_F - T_i)}$$

Note heat Q has dimension of energy, that is

$$[Q] = ML^2T^{-2},$$

and hence c has dimension of $[c] = L^2t^{-2}K^{-1}$. Since Q has units of J, the unit c of is $J(Kg)^{-1}K^{-1}$. The specific heat c of an object is a measure of how tightly bound together are its atoms (or molecules). For an object made out of tightly bound constituents, we expect c to be large. A lot of heat has to be supplied to it for increasing its temperature since a larger fraction energy would be stored in the form of potential energy instead of the kinetic energy of moving constituents, and vice versa for materials with loosely bound constituents.

Table. Specific Heat of Some Materials at Room Temperature

Material	$J(Kg)^{-1}K^{-1}$
Lead	128
Silver	236
Copper	386
Aluminum	900
Granite	790
Glass	840
Ice (278K)	2100
Mercury	140
Water	4190
Human Body (average)	3470

So, for example, we can say that metals must have low specific heat since they easily heat up, and this in turn implies that their electrons and atoms are not very tightly bound.

Ideal Gas

Any gas is specified by its temperature T, the volume it occupies V and its pressure P. Pressure is defined to be the force per unit area that the gas exerts on the walls which enclose its volume. T, P, and V are called state variables which specify the state of the gas. Pressure $[P]=ML^{-1}T^{-2}$ and volume $[V]=L^3$. Note that the state variables are intrinsic properties of the gas, and, unlike heat, do not depend on the history of how the gas arrived at these values. For example, in the phase diagram given above, the state variables say at volume V_0 and temperature T_0 shown above does not depend on the curve which resulted in the gas being at that point of the phase diagram; another process, at a different pressure, leads to the same value of V_0 and T_0, and it does not matter how the gas arrived at that point of phase space.

An ideal gas is a gas that is so dilute that its interactions can be completely neglected, and each atom of the gas can be said to behave as a free particle. A number of gases, including helium, oxygen, nitrogen and so on, do behave as ideal gases at moderate pressures and at moderate temperatures, that is temperatures that are far greater than their condensation temperatures. A derivation is given ofequation of state for the ideal gas based on the kinetic theory. The result is the following. Let N be the number of atoms of a gas in a volume V and the constant be the famous Boltzman's constant K given by,

$$k = 1.38 \times 10^{-23} JK^{-1}$$

We then have the ideal gas law given by,

$$PV = NkT$$

For a gas made out of the collection of molecules (composed of many atoms) the formula above is changed slightly. The state of an ideal gas given by. A phase diagram is one in which any two of its state variables, such as volume and pressure, are plotted as independent variables. A point in the phase diagram represents the state variables of the gas. An implicit assumption in drawing a phase diagram is that thermodynamic variables such as temperature, volume and pressure are continuous variables. This assumption is clearly reasonable, since, in an experiment, we can continuously vary these parametres. The phase diagram for and ideal gas at constant pressure yields a straight line of versus T and is shown in VT phase diagram in Figure . Similarly, the PV phase diagram shows how the pressure P and volume V is related for an ideal gas at different constant temperatures.

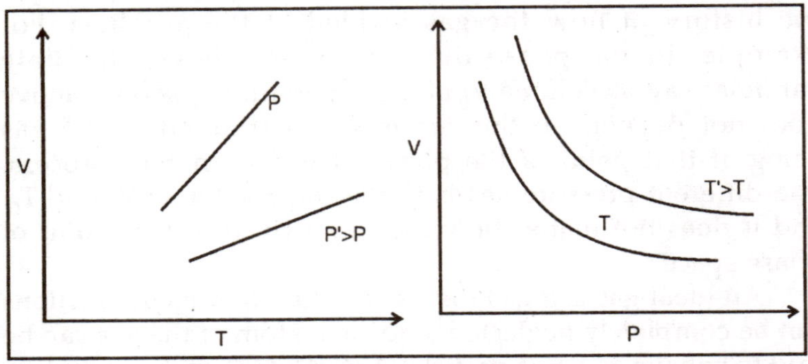

Fig. A VT and a PV Phase Diagram of a Gas

Amazing result. It states that equal volumes of a gas, at a given pressure and temperature, have the same number of atoms, regardless of the kind of atom it is made out of! And this conclusion, as shown, results directly from the application of Newton's Law to a large collection of atoms considered as classical particles. Suppose we hold the temperature of the gas at a constant value, and vary the pressure P. Let the initial values be , and the final values be ,P_f .V_f We then have from that,

$$P_iV_i = P_fV_f\text{: Boyle's Law}$$

Similarly, let us hold the pressure as fixed, and vary the temperature. Let initial values be v_i and T_i, and the final values be V_f and T_f.

Then we have,

$$\frac{V_i}{T_i} = \frac{V_f}{T_f}\text{: Charles's Law}$$

The internal energy U of an ideal gas is given by the following:

$$U = \frac{3}{2}NkT$$

Let us try and understand the physics of the above formulae. For a typical gas, occupying a volume of say 1 litre, the number of particles is close to Avogardo's number, which is given by $N_a = 6.02 \times 10^{23}$. Hence, the number of particles in any volume which we can directly perceive is truly unimaginably large.

The energy per particle is given by,

$$\frac{U}{N} = \frac{3}{2}kT$$

since energy of the particles is purely kinetic, we see that the result is in accordance with our earlier discussion that temperature is a measure of the kinetic energy of an atom in a gas.

The kinetic theory of gases is based on assuming that the position and velocity of the atoms are random. If the random velocity of an atom is v, then, the kinetic energy, , of the gas is simply the average kinetic energy U of the atoms.

Denoting the average of value of say v^2 by $< v^2 >$, we have, in three dimensional space.

$$\frac{U}{N} = \frac{1}{2}m < v^2 > = \frac{3}{2}kT$$

COMPLICATIONS WITH FIRST LAW

The main reason that, in the nineteenth century, physics shifted from the idea of force to the idea of energy was due to the need for explaining the phenomenon of heat.

If we include heat as a (rather unique) form of energy, the law of energy conservation then goes under the name of the First Law of Thermodyamics, which states the following.

Total Energy of System = Kinetic Energy + Potential Energy,
+ Heat
= Constant

Let us examine what form the First Law of Thermodynamics takes for the case of an ideal gas. We would like to apply force on the gas from outside, and determine how much energy and other properties of the gas are changed. We would also like to determine how much work we can extract from a gas.

The term work is used for indicating mechanical processes that involve moving around macroscopic bodies, such as pistons and pulleys.

The energy which is expended in doing such work is called mechanical energy, to differentiate it from the energy of the gas, which is called internal energy. Recall work results from a process of a force acting over some distance.

The sign convention for work is that work done by the gas is taken to be negative, whereas work done on the gas is taken to be positive.

In other words, a gas loses energy if it does work, whereas it will gain energy if work is done on it. Suppose the gas is contained in a frictionless piston of area , and we hold gas at a fixed temperature by keeping it in contact with a heat bath at temperature T; we do work on the gas by compressing its volume, so that $V_f<V_i$, so that $\Delta V = Vf-Vi<0$.

What is the work that the piston does on the gas? Let the piston exert a force F, and move through a distance Δx; the work done on the gas ΔW is given, as expected, by

$$\Delta W = F\Delta x$$

However, recall that the pressure P exerted by the gas is defined to be force per unit area.

Hence, we have,

$$F = PA$$

and consequently, since the volume is compressed $\Delta V = A\Delta x$, we have,

$$\Delta W = PA\Delta x$$
$$= P\Delta V$$

Suppose that heat of amount ΔQ flows into the gas during compression. We have, from the First Law of Thermodynamics (conservation of energy), the following. Change of Energy of Gas = Work done by gas + Heat flowing into gas,

$$\Delta E = -\Delta W + \Delta Q$$
$$= - P\Delta V + \Delta Q$$

The limit for infinitestimal change is given by,

$$dE = - PdV + dQ$$

Note ΔE is a function only of the state of the gas, and does not depend on how work was done on, or by, the gas, and how heat flowed in or out of the gas. However, the presence of ΔQ is process dependent, and we need further analysis to show that E is process independent.

Energy conservation does not tell us, for example, how much work can be transformed into heat, or whether all the heat in a body can be made to do "useful" mechanical work. The information about the direction in which energy can flow, and how much energy can be transformed from say heat into work, and vice versa, is given by the concept of entropy.

Entropy

We will have to discuss a number of concepts before we can precisely define what is the entropy - denoted by - of a physical system. It states that the fundamental result that we will arrive at, namely: For an isolated system, its entropy S can never decrease. The statement above tells us that all physical processes in which heat is transformed into work and visa versa must satisfy the condition that entropy must not decrease.

In other words, if this condition is not satisfied, the process cannot take place. For example a process in which there is a net decrease in entropy, that is, is forbidden as it would imply, as shown in Sec 2.6.3, that a perpetual machine could be constructed.

To define the concept of entropy quantitatively, we need to first discuss what is meant by a reversible and an irreversible process.

Consider a gas inside a frictionless piston in contact with a heat bath. If we pull out the piston very slowly, in a number of very small steps, the gas will expand at constant temperature , and if we then push the piston back, again slowly and in small steps, the gas will compress and return to its former state. This process is reversible as shown in Figure.

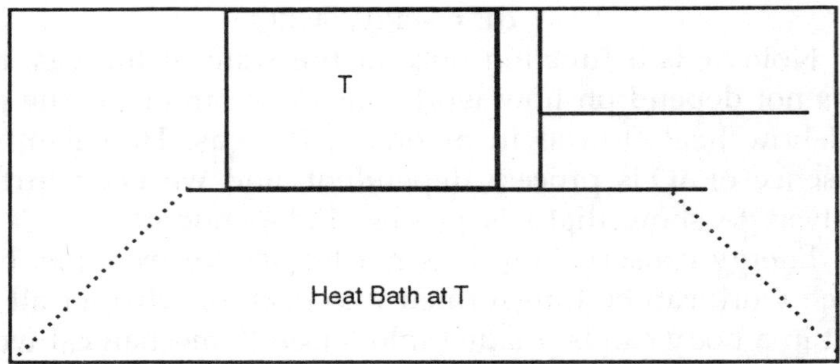

Fig. A Reversible Process

The point to note is that for a reversible process, for each step taken, the system is in equilibrium, and that a small step in one direction, say causing compression, can be exactly reversed by a small step in the opposite direction, causing expansion.

If we had suddenly pulled the piston out, the temperature of the gas would have dropped, there would have been turbulent gas motion, heat would have flowed into the gas from the reservoir and so on.

The process would then have become an irreversible one. No process in nature is fully reversible, and for a process to be considered as reversible, we need to ignore all effects such as friction, imperfect insulation and so on which would cause irreversiblity.

Idealization of a thermodyamic process as being reversible is analogous to Newton's idealization in stating his first law, since complications which come from friction and other irreversible effects are inessential in understanding the underlying principles of thermodyamics.

Consider a system in contact with a heat bath and held at constant temperature T; for a very small change in the independent parametres of the system such as volume, in which ΔQ amount of heat flows into it via a reversible process, the change in entropy of the system is given by,

$$\Delta S = \frac{\Delta Q}{T}$$

Say $[S] = ML^2T^{-2}K^{-1}$,

and the unit of entropy is JK^{-1}. For energy conservation we consequently have,

$$\Delta E = -\Delta W + \Delta Q$$

$$\Rightarrow \Delta E = -P\Delta V + T\Delta S$$

The entropy of an ideal gas is given by,

$$SG = Nk\ \text{In}(VT^{3/2}) + \text{Constant}$$

The entropy per particle (ignoring a constant) is hence given by,

$$S_G / N = k\,\text{In}(VT^{\frac{3}{2}})\cdot$$

We had assumed that for an ideal gas the interactions amongst the particles could be ignored. We see that the formula above confirms this, since the energy and entropy are both proportional to , which comes from a simple sum over the individual gas atoms. Although entropy is an intrinsic property of the system, it is not an intrinsic property of an individual particle, but rather depends, for example, on the total volume occupied by the gas.

To know what a particle is doing, we need to know both its position and velocity. We see that entropy is, in a precise sense, our degree of ignorance about the particle's behaviour. For example, entropy increases as we increase , which reflects the fact that the position of the particle becomes more uncertain, since it can be anywhere in a larger volume .

Similarly, if we increase temperature, entropy goes up since the larger the kinetic energy, the more random the velocity of the particle, and hence a greater degree in our ignorance about the particle.

REVERSIBLE PROCESSES

The most significant and remarkable feature of entropy is that, although ΔQ is a quantity which is process dependent, entropy -similar to energy- is anintrinsic property of the system, and unlike heat, entropy can be stored.

We hence have S = S(T,V) We examine what it means, for example, for energy and entropy to depend on only T and V.

Consider the VT-phase diagram. That is, consider T and V to be the independent variables, and let pressure P = P(T,V) be fixed by the ideal gas law. For every point on thisVT -diagram, both E and S have some definite value.

Suppose we take the ideal gas through any process, be it reversible or irreversible, along path 1, as indicated in Figure from values T,V to some new values T', V'. (Note this path is not a path in physical space, but rather a path in the phase space of T,V.) The system will go through a series of states with varying T,V, each having its own E,S.

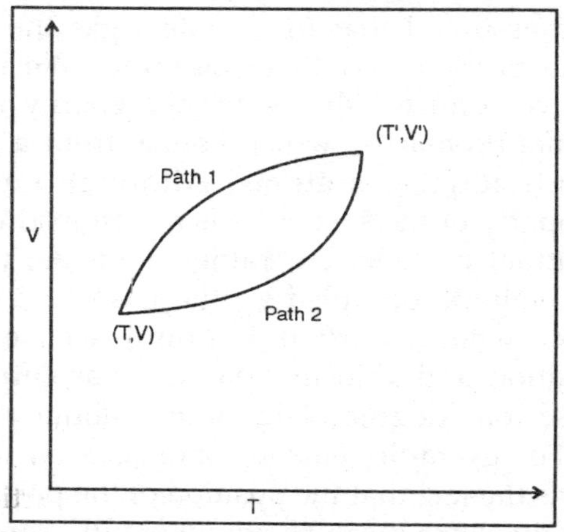

Fig. Phase Diagram of VT with Paths 1 and 2

The final change in energy and entropy is given by,

$$\Delta E = E(T',V') - E(T,V)$$

$$\Delta S = S(T',V') - S(T,V)$$

Now let us take the ideal gas through a different and arbitrary process following say path 2, also given in Figure Then ΔE and ΔS will have the same values as given in! In other words, as mentioned before, the values of E,S depend only on T,V, and not how the system arrived at these values. The change in the energy and entropy for a reversible process does not depend on the path taken; rather it only depends on the initial and final values of entropy.

If one goes through a closed path following a reversible process, it follows from above that the total change in the entropy of the system is zero, that is it is restored to its original state. However, note that in this reversible process, a net amount of heat has been expended. In symbols,

$$\Delta_S \text{ Closed Path } = 0 : \text{Reversible}$$

$$\Delta_Q \text{ Closed Path } = >0$$

Recall entropy depends only on the the state of the gas. Hence, if heat ΔQ is supplied to the gas through a reversible process, the total heat expended can be computed from the change in entropy caused by the heat flows.

Let the reversible process start with initial state of the gas being Ti, Vi and the final state of the gas being Tf, Vf as shown in figure. The total heat expended in the reversible process is,

$$Q_{Total} = \int_{Reversible}^{TDS}$$

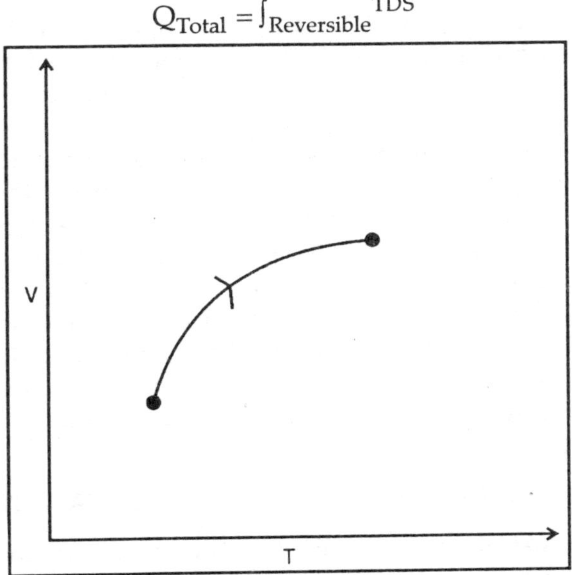

Fig. Reversible Process from Initial State to Final State

IRREVERSIBLE PROCESSES

Consider going around, at room temperature T, in a

circle in physical space. Since we always have to overcome friction for movement, certain amount of work W would be necessary to overcome friction.

From energy conservation this work would generate heat Q = W and lead to an increase of entropy,

$$\Delta_S = \frac{W}{T} > 0 .$$

Hence, in going around a closed path, we have not come back to the original state, since we have had heating due to friction, something that we can no longer reverse. In general, if we follow an irreversible process in completing a complete cycle, the entropy always increases.

That is,

$$\Delta_S \text{ Closed Path} > 0 : \text{Irreversible}$$

$$\Delta_Q \text{ Closed Path} > 0$$

How would one determine the change in entropy for an irreversible process? Since the formula we have for entropy given in eq. is valid only for reversible processes, we cannot use it.

We instead adopt the following approach. The irreversible process starts from some initial state of the system and ends up in a final state. We construct an equivalent reversible process which also goes from the same initial to the same final state.

Since entropy does not depend on how its values are attained, we can now compute the change of entropy for the equivalent reversible process, which will then be equal to the change in entropy for the irreversible process in question.

AN IRREVERSIBLE PROCESS:FREE EXPANSION

To illustrate how to calculate the change in entropy in an irreversible process, consider the irreversible free expansion of an ideal gas from an initial volume V_i into a vacuum such that its final volume is V_f. Suppose that the gas is thermally insulated from the environment, and at initial temperature of T_i.

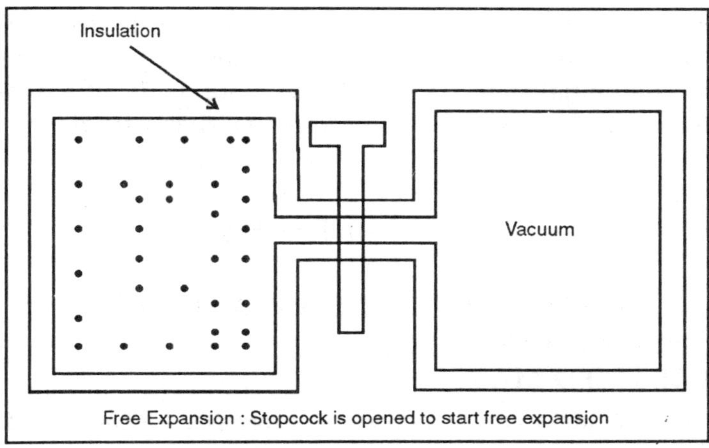

Fig. An Irreversible Process

Since the gas expands into a vacuum it does no work, that is $\Delta W=0$; furthermore, since it is thermally insulated there is no flow of heat, and hence $\Delta Q=0$.

The first law implies that,

$$\Delta_E = -\Delta W + \Delta Q$$
$$= 0$$

Since total energy does not change for an ideal gas, this implies that the initial and final temperature T_f are equal.

Hence,

$$T_i = T_f = T$$

So the initial state of the gas is $T_i = T$, V_i and the final state of the gas is $T_f = T$, V_f. We would like to replace the irreversible free expansion of the ideal gas with an equivalent reversible process. Since the initial and final temperatures are the same, we can replace the free expansion with an (isothermal) expansion of the gas at constant temperature, from volume V_i to V_f. To see that expansion at constant temperature T is a reversible process, consider a gas in a piston and in contact with a heat bath at temperature T. If one pulls the piston very slowly, heat will flow in to keep the temperature at ; on the other hand, if one slowly compresses the piston, the gas will be compressed, and heat will flow back into the reservoir. So we can see that an isothermal expansion is reversible. All we need to do now is to compute the difference between the entropy of an ideal gas in the initial state T_i, Vi and final state T_f, V_f. We hence have the total change in the entropy as given, from, by

$$\Delta_S = S_G(T, V_i) - S_G(T, V_f)$$

$$= Nk\ In\left(\frac{V_f}{V_i}\right) > 0 : \text{Free Expansion}$$

CHANGE OF ENTROPY OF GASES

ENTROPY CHANGE FOR IDEAL FIXED MASS

For an ideal constant volume process... ($dQ = mc_v dT$)

$$dS = \frac{dQ}{T} = \frac{mc_v dT}{T}$$

Integrating between initial (2) and final state (2),

$$S_2 - S_1 = mc_v \log_e \frac{T_2}{T_1}$$

For an ideal constant pressure process... ($dQ = mc_p dT$),

$$dS = \frac{dQ}{T} = \frac{mc_p dT}{T}$$

Integrating between initial (2) and final state (2),

$$S_2 - S_1 = mc_v \log_e \frac{T_2}{T_1}$$

For an ideal adiabetic process...

$$dQ = 0$$

$$dS = \frac{dQ}{T} = 0$$

For an ideal isothermal process...

There is no change in temperature and therefore there is no change in internal energy U. From the first law of thermodynamic $dQ = dU + dW$. If $dU = 0$ then $dQ = dW$.

$$dS = \frac{dW}{T} = \frac{pdV}{T}$$

$$\text{as } pV = mRT \text{ or } T = \frac{pV}{mR}$$

$$dS = \frac{pdvmR}{pV} = \frac{mRdV}{V}$$

ENTROPY CHANGES IN TERMS OF PROPERTIES OF A PERFECT GAS

From the definition of entropy i.e $\delta Q_{rev} = T\delta S$ and from the first law of thermodynamics is dQ = dE + dW which, for a reversible process can be written as $\delta Q_{rev} = \delta E + P\delta V$... it can be deduced that dE = TdS - PdV for a reversible process and if the substance is a gas of mass m then,

$$dQ = TdS = mC_v dT + PdV \text{...........equation x}$$

For a perfect gas PV = mRT and assuming C_v is constant.

a) Entropy change in terms of V and T

By substituting,

$$P = \frac{RmT}{V}$$

in equation x,

$$\int_1^2 dS = mC_v \int_1^2 \frac{dT}{T} + Rm \int_1^2 \frac{dV}{V} \quad \text{or}$$

$$S_2 - S_1 = mC_v \log_e\left(\frac{T_2}{T_1}\right) + Rm \log_e\left(\frac{V_2}{V_1}\right)$$

b) Entropy change in terms of P and T

By differentiating PV = mRT,

$$PdV = RmdT - VdP = RmdT - \frac{RmT}{P}dP$$

Substituting above in x and integrating to get,

$$\int_1^2 dS = m(C_v + R)\int_1^2 \frac{dT}{T} - Rm \int_1^2 \frac{dP}{P} \quad \text{or}$$

$$S_2 - S_1 = mC_P \log_e\left(\frac{T_2}{T_1}\right) - Rm \log_e\left(\frac{P_2}{P_1}\right)$$

c) Entropy change in terms of P and V,

$$T = \frac{PV}{Rm}$$

and differentiating PV = mRT

Result in,

$$dT = \frac{PdV + VdP}{Rm}$$

Substituting above in x and integrating to get,

$$\int_1^2 dS = m(C_v + R)\int_1^2 \frac{dV}{V} - mC_v \int_1^2 \frac{dP}{P} \quad \text{or}$$

$$S_2 - S_1 = mC_P \log_e\left(\frac{V_2}{V_1}\right) - mC_v \log_e\left(\frac{P_2}{P_1}\right)$$

EXAMPLE

To determine the entropy change in the irreversible adiabetic expansion of Joules experiment....

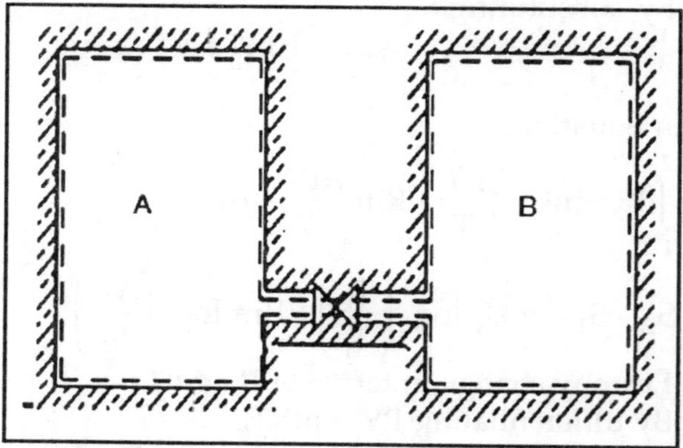

In this experiment a gas at high pressure (properties P_1, V_1,T)in chamber A which is connected to a second evacuated chamber B via a valve. When the valve is opened the gas in chamber A moves by free unresisted expansion into chamber B until the pressures equalise (final properties P_2, V_2,T). The gas is supposed perfect and the temperature does not change (no work has been done). An irreversible expansion has taken place. The equipment is so arranged that there is no heat transfer between the surroundings and the system ($\delta Q = 0$). Although $\delta Q = 0$ the process is irreversible and therefore δQ is not zero. To determine the change in entropy the process is replaced by a reversible process which has the same initial and final conditions i.e. an isothermal process.

For an isothermal process the change in entropy,

$$= S_2 - S_1 = mR\log_e \frac{V_2}{V_1}$$

In this case although the process is adiabetic because it is not reversible it is not isentropic. There is a resulting increase in entropy reflecting the various irreversible energy losses in this process.

ADIABETIC,ISENTROPIC, AND ISOTHERMAL PROCESSES

Ideal reversible adiabetic and isothermal processes are straight lines on the TS chart as shown above ac = ideal

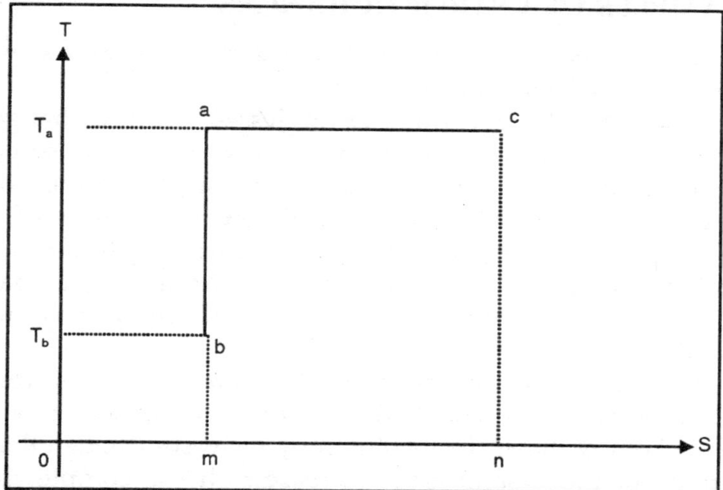

isothermal process and ab = ideal isentropic process. An adiabetic process is one which is insulated against heat transfer and an isentropic process is one with a constant entropy.

δQ may be zero during a adiabetic process but if the process is not reversible then $\delta S = \delta Q/T$ does not apply and δS is not zero. If the process is reversible then $\delta Q = 0$ and $\delta S = 0$ i.e the process is adiabetic and isentropic.

A isentropic process is not necessarily reversible for a real expansion or compression-with all the associated eddies and friction associated with real processes may have sufficient heat transfer to maintain the entropy of the relevant substance constant ($\delta S = 0$). A real process could be isentropic but would not at the same time be adiabetic because some heat transfer would be necessary. The following relationships are possible

- In a real (irreversible) adiabetic process dQ = 0 but dS > 0
- In a reversible adiabetic process dQ = 0 and dS =0 0
- In a real (irreversible) isentropic process dS = 0 but dQ < 0
- In a reversible isentropic process dS = 0 and dQ = 0 0

EXAMPLES OF ENTROPY CHANGES

Take-home message: Spontaneous changes are aways associated with entropy increase.

Below we have various examples of entropy change during various processes. From them we can draw some general conclusions. First we see in general the entropy of anything increases when it is heated, and the entropy of a gas increases when it expands at constant temperature. A common theme (to be explored in more detail later) is that an increase in entropy is associated with an increase in disorder.

Examples of spontaneous processes are the flow of heat from a hotter to a colder body and the free expansion of a gas. In both of these, the total entropy increases (though that of parts of the system may decrease). This is completely general: spontaneous processes are those which increase entropy. Since heat flow from hotter to colder bodies is irreversible, reversible processes must involve heat flow between bodies of the *same* temperature. It follows that any entropy change of the system must be exactly balanced by that of the heat bath which provided the heat: . Thus the entropy change of the universe during reversible processes is zero. During an adiabatic process no heat flows. Thus from we see that the entropy change of a 'system during for a reversible, adiabatic process is zero. But note that both qualifiers are needed; the entropy of a non-isolated system can change during a reversible process (and the entropy change of the surroundings will compensate), and an irreversible change to an isolated system will increase the entropy.

Try these problems for yourself before checking the detailed answers!

- *Example:* Two identical blocks of iron, one at 100°C and the other at 0°C, are brought into thermal contact. What happens? What is the total entropy change? (Assume the heat capacity of each block, C, is constant over this temperature range, and neglect volume changes)

Answer: Both blocks end up at 50°C and the entropy change is 0.024C.

Two identical blocks of iron, one at 100°C and the other at 0°C, are brought into thermal contact. What happens? What is the total entropy change? (Assume the heat capacity of each block, , is constant over this temperature range, and neglect volume changes)

We know what happens: heat flows from the hot to the cold body till they reach the same temperature; conservation of energy requires that this will be at 50°C. Why does heat transfer occur? If heat $đQ$ is transferred from a hot body at T_H to a cold one at T_C, the entropy decrease of the hot body is $dS_H > -đQ/T_H$, and the entropy increase of the cold body is $dS_C > đQ/T_C$. So the total entropy change is,

$$dS = dS_H + dS_C > \left(\frac{1}{T_C} - \frac{1}{T_H}\right) đQ > 0$$

So the decrease in entropy of the hot block is more than compensated by the increase in entropy of the cold block. The spontaneous flow of heat is associated with an overall entropy increase, and the two blocks exchange heat till their combined entropy is maximised.

What is the overall entropy change for the total process?

Here we have to use a trick. We can't calculate entropy changes for irreversible processes directly; we need to imagine a reversible process between the same endpoints. For the heating or cooling of a block, this would involve bringing it in contact with a series of heat baths at infinitesimally increasing or decreasing temperatures, so that the temperature difference between the heat bath and the block is always negligible and the entropy change is zero.

During this process, the heat transfer and the infinitesimal temperature change are related by,

$$đQ^{rev} = CdT$$

$$\Delta S = \int_1^2 \frac{đQ^{rev}}{T} = C\int_{T_1}^{T_2} \frac{dT}{T} = C \ln\left(\frac{T_2}{T_1}\right)$$

Thus the total entropy change is,

$$\Delta S = \Delta S_C + \Delta S_H = C \ln\left(\frac{T_f}{T_C}\right) + C \ln\left(\frac{T_f}{T_H}\right) = C \ln\left(\frac{T_f^2}{T_H T_C}\right)$$

$$= C \ln\left(\frac{323^2}{273 \times 373}\right) = 0.024C$$

- *Example:* Two identical blocks of iron, one at 100°C and the other at 0C, are brought into thermal contact. What is the maximum work that can be extracted from the hot block in the absence of other heat sinks?

Answer: The final temperature is 46°C and 14% of the energy lost by the hot block is available to do work.

Two identical blocks of iron, one at 100°C and the other at 0°C, are brought into thermal contact. what is the maximum work that can be extracted from the hot block in the absence of other heat sinks?

Remember, we can't just extract heat from the hot block and turn it into work; the entropy of the block would decrease without any compensating increase elsewhere. We need to add at least enough heat to the cold block so that its entropy increases by as much as that of the hot block decreases. (Work can always be used to do things which don't increase the entropy of the universe, such as lifting a weight.) Once the two blocks are at the same temperature, no further work can be extracted. At that point the entropy change of the two blocks together, from the previous example, is

$$\Delta S = c \ \text{In}\left(\frac{T_f^2}{T_H T_C}\right) \geq 0$$

The lowest final temperature is that for which , *ie*

$$T_f = \sqrt{T_H T_C} = 319 \text{ K} = 46°\text{C}.$$

(Any lower and ΔS would go negative, which isn't allowed.) So although the total heat loss of the hot block is $Q_H=c(T_H-T_f)$, the work extracted is only the difference between this and the heat gained by the cold block, $Q_C=c(T_f-T_C)$, namely $W=c(T_H+T_C-2T_f)$.

In this case, the efficiency is,

$$\eta = \frac{W}{Q_H} = \frac{(T_H + T_C - 2T_f)}{(T_H - T_f)} = 0.144$$

- *Example:* 3 Heat engines revisited From the law of non-decrease of entropy, show that the maximum efficiency of a heat engine operating between two reservoirs at and occurs when the engine is reversible.

Answer: Heat engines revisited: From the law of non-decrease of entropy, show that the maximum efficiency of a heat engine operating between two reservoirs at T_H and T_C occurs when the engine is reversible.

This is a bit circular, as we used the properties of Carnot engines to derive this form of the second law! However we will see later in the course that it can be independently derived using statistical methods. The change in entropy of the two reservoirs, which must be non-negative, is

$$\Delta S = \Delta S_C + \Delta S_H = \frac{Q_C}{T_C} - \frac{Q_H}{T_H} \geq 0 \Rightarrow \frac{Q_C}{Q_H} \geq \frac{T_C}{T_H}$$

with the equality being for ΔS=0, *ie* for a reversible process. So the efficiency is,

$$\eta = \frac{W}{Q_H} = 1 - \frac{Q_C}{Q_H} \leq 1 - \frac{T_C}{T_H}$$

This is maximum when the equality is satisfied, *ie* for a reversible engine.

- *Example:* n moles of an ideal gas at temperature T_0 are originally confined to half of an insulated container by a partition. The partition is removed without doing any work. What is the change in entropy?

Answer: ΔS=nRIn2. More details n moles of an ideal gas at temperature T_0 are originally confined to half of an insulated container by a partition. The partition is removed without doing any work. What is the final change in entropy?

First, we must resist the temptation to say $đQ = 0$ (because process is adiabatic) hence dS=0. In fact $dS = đQ^{rev}/T$ and this is not a reversible process. To solve this problem, we need to find a reversible process linking the same two endpoints. We've already seen this; the process is a reversible isothermal expansion

For such an expansion,

$$Q= -W=nRT_0 \text{ In}(V_2/V_1) \text{ so } DS=Q/T_0=nR \text{ In}2.$$

This is rather subtle; make sure you understand the difference between the actual process (with Q=0 and W=0) and the reversible process for which we could calculate the entropy change (with Q= -W≠0). Remember, Q and W are not functions of state - only the sum is.

- *Example:* An insulated container is originally divided in half by a partition, and each half is originally occupied by n moles of an different ideal gas at temperature T_0. The partition is removed without doing any work. What is the change in entropy?

Answer: DS=2nR In2. An insulated container is originally divided in half by a partition, and each half is originally occupied by moles of an different ideal gas at temperature . The partition is removed without doing any work. What is the change in entropy?

Since these are ideal gases, they do not interact. Each species is oblivious of the existence of the other and the entropy change for each is just the same as in the free expansion of . Thus the total entropy change is given by . Note that the total mixing - which we know will happen eventually - is exactly the change which maximises the entropy.

Chapter 4

Vapouer, Gas and Steam

GAS AND VAPOUR

DIFFERENCE BETWEEN GAS AND VAPOUR

Vapour is a substance which has experienced a phase change, whereas gas is a substance which has not, and will not experiece a phase change.Vapour is a substance which is near the condensationGas is very far away from condensation.Below the critical temperature of a substance, it is known as a vapour. It can be liquified. But, a gas exists after the critical temperature. It can not be liquified.The word vapour in its natural state is a solid or liquid at room temperature.

However, a gas in its natural state at room temperature would still be a gas.

Example:

1) Steam would be a vapour because at room temperature, it would be water, which is a liquid.
2) Nitrogen (a gas) at room temperature would still be in a gaseous state. Yes, there is a difference.

To make it more simple, a vapour is a substance which has experienced a phase change. Whereas, a gas is a substance which has not, and will not experience a phase change. Vapour is a Substance which is near the condensation Gas is very far away from condensation In other word: Gas is something that u cannot make it to liquid Only by increasing the pressure on it!! U have to decrease the tempreture aswel, otherwise it wouldnt turn into liquid!Basic difference is gas is a completely state of matter while vapour is not a state of matter bcz it contains liquid content.hope that answer the Question.

VAPOUR, GAS, STEAM, FLUID

- There is no significant physical or chemical difference between a vapour and a gas. This is the most important point.
- However, the words have slightly different connotations. Remember that these are just connotations. There is sometimes considerable overlap between the terms, so precise distinctions are not necessary and probably not even possible.
- Gas is the broader term. A vapour is a particular type of gas. Any vapour could *technically* be called a gas, although in many situations this would sound weird or awkward.
- There seem to be two main ways in which a gas can earn the "vapour" appellation:
  - When the gaseous phase is in equilibrium with the corresponding liquid or solid, or
  - When we wish to emphasize that we are talking about the gaseous phase, even though for some reason the condensed phase is considered "normal".
- Here are some examples, with comments.

Thoughtful experts refer to gaseous H_2O as "water vapour" even in situations where it is nowhere near being in equilibrium with any condensed phase. Meteorology is one example, where we speak of various amounts of water vapour in the atmosphere. Astronomy is an even more extreme example, where congregations of cosmic H_20 molecules are called interstellar water vapour. Since the dawn of time, these particular H_2O molecules have been nowhere near saturated vapour pressure. Nobody ever speaks of water gas.

This is the perfect illustration of rule (b). Technically and officially, "water" refers to the H_2O molecule in any phase (gas, liquid, or solid)... but unofficially by long tradition "water" connotes the liquid unless otherwise specified. Water is*normally* a liquid.

- If you smell fumes in the garage, you call it gasoline vapour, not gasoline gas.

This is another illustration of rule (b). Unofficially, gasoline is normally presumed to be a liquid unless otherwise stated. Therefore the gaseous phase is called vapour, even in situations where the partial pressure is nowhere near the saturated vapour pressure.

- You can fill balloons from a tank of helium gas. It would sound quite weird to call it helium vapour.

The gas phase is considered normal for helium, and in this case there is no liquid or solid anywhere in sight.

- In contrast, you can produce temperatures below 2K if you start with liquid helium and pump the vapour away fast enough.

This is the perfect illustration of rule (a). The gas is called vapour to hint at the presense of the nearby liquid.

- Thoughtful experts commonly refer to the "SVP" (Saturated Vapour Pressure) to designate a vapour that is in equilibrium with a nearby condensed phase.

This proves that rule (a) cannot be the whole story; otherwise the word "Saturated" would be redundant.

- It has been suggested that a vapour is something that you can see or smell, or something that settles to the ground – but that's can't possibly be right. Ammonia (NH_3) is in many situations called ammonia gas, despite its pungent odor. Water vapour is called vapour, even though it is odorless and highly transparent. Fog and clouds are visible but are neither gases nor vapours; they are aerosols, i.e. colloidal suspensions of very fine particles of liquid (or sometimes solid) H_2O in the air. The aerosol as a whole has the property of filling its container, so it behaves like a gas in this regard. The suspended particles will not settle out. (A system where the droplets are so big that they do eventually settle out should be called a mist.)

- The word "steam" applies both to transparent water vapour and to visible clouds of water suspended in air.
- The word "fluid" is a very important technical term. It encompasses both gas and liquid. That is, any gas is a fluid and any liquid is a fluid. Above the critical temperature and/or critical pressure, "fluid" is the preferred term, because there is no meaningful distinction between liquid/vapour/gas. But this is not an if-and-only-if proposition; the term fluid may perfectly well be applied to liquids and gases at temperatures and pressures below T_c and P_c

VAPOUR RECUPERATOR

According to the invention, an increase in the coefficient of performance (COP) and specific refrigeration capacity is provided by a vapour recuperator, comprising a heat exchanger located along the flow paths of the gaseous reactant to and from the reactor(s). As illustrated in FIG. 1, the vapour recuperator 40 is placed conveniently along the conduits 38 and 39 between reactors 10 and 20 and the evaporator 30 and condenser 32, respectively. At such a location, the recuperator 40 provides for heat exchange between the gaseous reactant vapour streams flowing between the reactor(s) and the condenser, and between the evaporator and the reactor(s).

The recuperator may be located on either side of the valve 35, although where check valves are used, the position illustrated is preferred. By incorporating such a recuperator, super-heated vapour flowing from the desorption reactor(s) toward the condenser is cooled against the relatively cool vapour directed from the evaporator to the adsorption reactor(s). Because energy recuperated from the superheated refrigerant leaving the desorption reaction is transferred to the cold gaseous refrigerant typically leaving the evaporator and then undergoing exothermic adsorption, thermal efficiency of the system is increased.

In the embodiment shown in Fig. reactors 12 and 14 are substituted for the evaporator and condenser components used in the refrigerant phase charge apparatus. Such reactors, contain a solid or liquid salts for alternately adsorbing (absorbing) and desorbing the gaseous reactant directed thereto from the staging reactors 10 and 20. The reactors 12 and 14 cooperate with heat exchanges for recovery of energy from the alternating chemisorption reaction as described in the aforesaid application and incorporated herein by reference.

The vapour recuperator 40 functions the same way in this embodiment as in Fig. to cool super-heated refrigerant vapour directed from a staging desorbing reactor (10 and 20) to an adsorbing reactor (12 or 14), against the relatively cool vapour directed from a desorbing reactor (12 or 14) to a staging adsorbing reactor (10 or 20).

CHANGE OF PHASE DURING CONSTANT PRESSURE PROCESS

Phase Changes

That when a system reaches an unstable state it splits itself into different phases with different properties but in equilibrium with each other. was just a halt to change our view-point and see how the balance equations change when we let mass flows through the frontier of our system. Now we proceed with the thermodynamics of simple compressible system, and we analyse phase changes in a pure chemical substance.

Only phase changes in pure small-molecule substances are considered (i.e. usual gases, water and other simple inorganic liquids, small-chain hydrocarbons, metals and other inorganic solids), but not glasses, polymers and composites. It is interesting, however, to look at how, for instance, honey (or milk, oil, wax) boils and solidifies, or at how wax melts (or paper, wood, teflon, concrete, and so on). Furthermore, we focus on the liquid/vapour phase change, covering other phase changes occasionally; solid/solid allotropic transformations are not mentioned at all. Glasses are substances (inorganic as silicates or organic as plastics) that solidify from a molten state into an amorphous solid (i.e. without crystalline order);

It is also important to mention that although we only deal here with phase changes of aggregation in a pure substance other thermodynamic phase changes occur in Nature, as the change from normal electrical conductivity to superconductivity (with or without magnetic field), the change from paramagnetism to ferromagnetism, the change from normal viscosity to superfluidity in ^{4}He, etc. The second phase transition discovered was the ferromagnetic-paramagnetic transition at high temperature, when J. Hopkinson described in 1890 the disappearance of magnetism in hot iron. Later on, electrical superconductivity was discovered in 1911 by Kamerlingh Onnes, in mercury.

We only deal with phase changes at equilibrium conditions, and do not consider the kinetics of the change, dominated by mass diffusion and nucleation processes.

Besides, it must be mentioned that there is a whole field in Thermodynamics, not covered here, that deals with the critical region, i.e. the neighbourhood of critical points. The existence of a critical temperature above which a gas cannot be condensed by increasing pressure was first pointed out in 1822 by Cagniard de la Tour, studied in detail by D. Mendekeyev in 1860, and fully described by T. Andrews in 1869 in terms of the isothermal behaviour with pressure of carbon dioxide, at several temperatures; three years later, van der Waals proposed the first embracing theory for phase transition, with his famous state equation modelling both the gas and the liquid states.

PHASE RULE

The condition of equilibrium for a pure substance introduced three intensive variables T, p and $\grave{\imath}$ (the chemical potential of the substance), but they were not all independent since the Gibbs-Duhem equation, $0=sdT-vdp+d\grave{\imath}$, correlates them, thus a simple thermodynamic system has only two independent intensive variables. When there are several phases, the same equilibrium variables hold (T, p and $\grave{\imath}$), but now there is a Gibbs-Duhem equation for each of the phases, since each one has different *s* and *v*. Thus, the variance (intensive degrees of freedom) of a single phase is 2 (e.g. *T* and *p*), that of a two-phase system is 1 (only *T* or *p*), and that of a three-phase is 0 (it is a constant state without any possible variation of *T* or *p*), not being possible a four-phase system (for a pure substance).

CLAPEYRON EQUATION

Clapeyron equation gives the relation between *T* and *p* in a two-phase pure-substance system. It is deduced by just subtracting the two Gibbs-Duhem equations for the two phases at equilibrium. Using the notation for the liquid-to-vapour phase change, the most important in Thermodynamics, one gets:

$$\left.\begin{array}{l} 0 = S_V dT - v_v dp + d\mu \\ 0 = S_L dT - v_L dp + d\mu \end{array}\right\} 0 = (S_V - S_L)dT - (V_V - V_L)dp$$

$$\Rightarrow \left.\frac{dp}{dT}\right|_{sat} = \frac{h_V - h_L}{T(v_V - v_L)} = \frac{h_{LV}}{T_{vLv}}$$

After Δs has been substituted by $\Delta h/T$ based on the general relation (4.2), $dh = Tds + vdp$, and the fact that both equilibrium states (liquid and vapour) have the same pressure.

The enthalpy change in a phase change, e.g. $h_{LV} = h_V"h_L$ for vapourisation, was initially named 'latent heat' by J. Black. Notice that subindex 'sat', meaning 'saturated' or better 'in saturation', is used as a synonym of biphasic equilibrium, and that the loci of liquid-vapour equilibrium states is known as vapour pressure curve, being bounded by the triple point and the critical point.

Equation below is the general Clapeyron equation and to go any further we need to have a model for the substance. An integrated Clapeyron equation is obtained if one makes the following three simplifications: $v_L \ll v_V$, $v_V = RT/p$ and h_{LV}=constant, what yields:

$$\Rightarrow \left.\frac{dp}{dT}\right|_{sat} = \frac{h_V - h_L}{T(v_V - v_L)} \xrightarrow{v_V \gg v_L,\, v_V = RT/p,\, h_{Lv=const}}$$

$$\ln\left(\frac{p}{p_0}\right) = \frac{-h_{Lv}}{R}\left(\frac{1}{T} - \frac{1}{T_0}\right)$$

$$\Rightarrow \left.\frac{dp}{dT}\right|_{sat} = \frac{h_V - h_L}{T(v_V - v_L)} \xrightarrow{v_v >> v_L, v_v - RT/p, h_{Lv} - const}$$

$$\ln\left(\frac{p}{p_0}\right) = \frac{-h_{LV}}{R}\left(\frac{1}{T} - \frac{1}{T_0}\right)$$

which may be used if one saturated point is known (T_0-p_0; e.g. the normal boiling point, T_b, at p_0=100 kPa) to compute some neighbouring saturated state, far from the critical region.

Clapeyron equation is often known as Clausius-Clapeyron equation; Émile Clapeyron, a co-worker of Carnot, proposed it around 1834 to find the variation of vapourisation enthalpy with temperature, and Rudolph Clausius reformulated it in the 1850s. The name 'Clapeyron equation' is sometimes ascribed to the ideal gas law $pV=mRT$. The p-V diagram is occasionally named Clapeyron diagram.

Based on the functional dependence of vapour pressure with temperature given by (6.2), and trying a best fit to experimental data, Charles Antoine proposed in 1888 an empirical fit that is used the most:

$$\ln\left(\frac{p}{p_0}\right) = A - \frac{B}{\frac{T}{T_0} + C}$$

with A, B and C tabulated usually for T_0=1 K and p_0=1 kPa.

A table with these 3 constants for usual pure substances is presented in Thermal data tables. Finally, it is worth rewriting the general Clapeyron equation in term of reduced variables (dividing T-p values with those at the critical point), obtaining:

$$\left.\frac{dp}{dT}\right|_{sat} = \frac{h_v - h_L}{T(v_v - v_L)} \Rightarrow \frac{d\ln p_R}{d\frac{1}{T_R}} = \frac{h_v - h_L}{RT_{CR}(Z_V - Z_L)}$$

It is amazing that for many substances the slope of the vapour pressure curve in the $\ln p_R$-$1/T_R$ diagram is constant, from the triple point to the critical point; it is predicted by the integrated Clapeyron equation but that is only valid locally and far from the critical point. What happens is that the enthalpy of vapourisation (the numerator in varies with temperature as the difference in compressibility factor does, and both effects compensate.

There are some other similar generic approximation rules worth mention. Guldberg's rule points out that the ratio of boiling to critical temperatures for most substances is around T_b/T_{CR}0.6. Trouton's rule points out that $h_{lv}/(RT_b)$10, which can be deduced from the two precedent rules. Watson rule establishes a generic approximation to the variation of vapourisation enthalpy with temperature,

$h_{lv}/h_{lv,b}$»[(1"T_R)/(1"T_b/T_{CR})]$^{0.38}$ »[2.5(1"T_R)]$^{0.38}$.

Statement

Approximate the vapour-pressure curve for water by the following methods (given the values for 120 °C):

- By the reduced equation ln*p*R=*K*(1-1/*T*R), finding *K* by using the corresponding states model and a one-point fitting at *T*R=0.75.
- By the integrated Clapeyron's equation fitted at the normal boiling point.

$$\text{In}\left(\frac{p}{p_0}\right)=\frac{-h_{LV}}{R}\left(\frac{1}{T}-\frac{1}{T_0}\right)$$

- By Antoine's equation.

$$\text{In}\left(\frac{p}{p_0}\right)=A-\frac{B}{\dfrac{T}{T_0}+C}$$

Se trata de comparar varios métodos para determinar la curva de presión de vapour del agua, y en particular calcular la presión de vapour a 120 °C. Se pide:

- Aproximar la curva de presión de vapour del modelo de estados correspondientes mediante la ecuación ln*p*R=*K*(1-1/*T*R), determinando *K* por ajuste en el punto *T*R=0,75.
- Utilizando la ecuación integrada de Clapeyron a

partir del punto de ebullición normal.

- Utilizando la ecuación de Antoine.

Solution

- By the reduced equation ln*p*R=*K*(1-1/*T*R), finding *K* by using the corresponding states model and a one-point fitting at *T*R=0.75. Looking at the Z-p_R diagram (in Thermal Data), for *T*R=0.75 we find *p*R=0.11, and from the given equation *K*=6.6. Thus, for *T*=120 °C,*T*R=393/647=0.61, *p*R=exp(6.6(1-1/0.61))=0.013 and *p*=0.30 MPa.
- By the integrated Clapeyron's equation, fitted at the normal boiling point.
  It is with p_0=100 kPa, T_0=373 K (100 °C) and h_{LV}=2.26 MJ/kg. Thus, for *T*=120 °C, *p*=0.195 MPa.
- By Antoine's equation.
  It is with *A*=16.54, *B*=3985 and *C*=-39 for T_0=1 K and p_0= 1 kPa. Thus, for *T*=120 °C, *p*=0.197 MPa.

Comments

Roughly, the vapour pressure for water at 120 °C is 200 kPa; Antoine equation is the most used, Clapeyron equation is not bad so close to the normal boiling point, but the corresponding states model for water is rather inaccurate (water is one of the worst-behaved substance for 'generalised' models). Notice the different apparent curvature of the vapour pressure curve in the *p*-*T* and ln(*p*)-*T* diagrams.

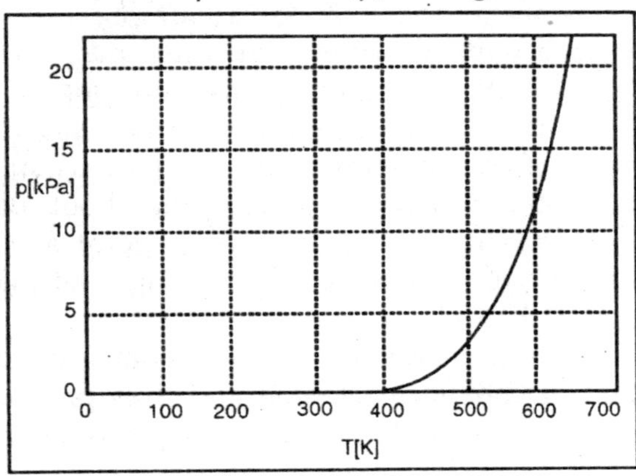

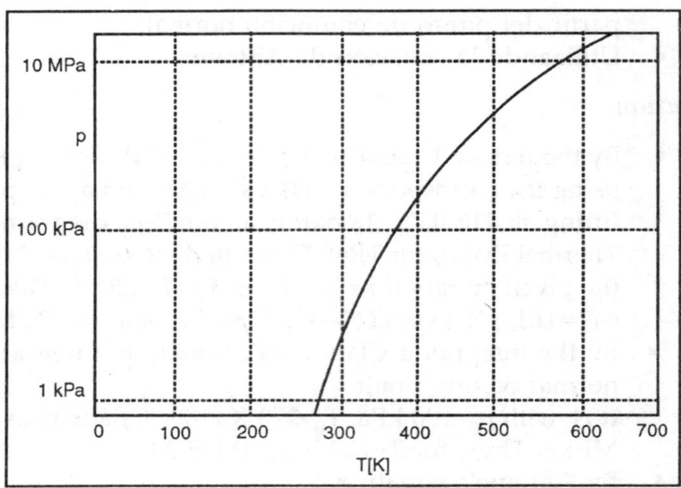

Clapeyron equations may also be applied to the solid-vapour phase change (sublimation) by changing the values of the variable's accordingly.

Substances with sublimation points are not common, the best known being carbon dioxide (CO_2, with T_{subl}=195 K), called dry ice, and less known acetylene (C_2H_2, withT_{subl}=189 K), sulfur hexafluoride (SF_6, with T_{subl}=209 K), uranium hexafluoride (UF_6, with T_{subl}=329 K), arsenic (As, with T_{subl}=887 K), carbon (C, graphite, T_{subl}=3910 K), and some mercury compounds.

Notice, however, that the strict meaning of sublimation-point used here, i.e. the solid-to-vapour phase-change of a pure substance with its triple-point pressure below normal pressure, p_0=100 kPa, is rarely followed not only in common language but in scientific literature, where sublimation is synonymous of evaporation from a solid, usually in the presence of ambient air, and thus, one talks about sublimation of iodine, I_2, having a normal boiling point at T_b=387 K, sublimation of naphthalene, $C_{10}H_8$, having a normal boiling point at T_b=491 K, etc.

Evaporation of solids in air is a general rule, forced by the gradient of chemical-potential and thus proportional to its vapour pressure, but their vapour pressure at normal temperatures is very low and, although all solids sublimate, their sublimation is usually neglected except in the examples given. Water ice sublimates too, below freezing temperatures, as in a freezer.

For the solid-liquid phase change, the slope of the equilibrium curve in the *p-T* diagram is so high that it is usually assumed that the temperature of fusion is the same at every pressure (this is used to approximate the triple point by the normal melting-point temperature and its corresponding vapour pressure).

However, it can be seen from the Clapeyron equation that the normal slope is positive because $h_L"h_S$ is always positive and $v_L"v_S$ is usually positive (substances contract on freezing), although the slope is negative for substances that expand on freezing, as water, silicon, germanium, gallium, bismuth, mercury, and others; in the case of pure water, it takes 13.4 MPa to freeze at –1°C (the minimum equilibrium freezing temperature is –22 °C, at 210 MPa, but lower liquid temperatures can be obtained even at normal pressure under metastable conditions).

By the way, the high pressure under an ice-skate blade is far too short to melt the ice (held at "7.."2 °C); the lubrication is due to a microscopic liquid layer at equilibrium on the ice surface (10^{-10}..10^{-7} m thick, depending on temperature and composition), greatly enhanced by the local fusion caused by friction.

Phases and phase changes for a pure substance are named as in, jointly with a three-dimensional view of equilibrium states, or *p-v-T* diagram; i.e. the traditional solid, liquid and gas phases, and the phase change from one to other. Allotropic changes, in the solid state, are not considered. The supercritical region is sometimes limited to $T>T_{CP}$ and $p>p_{CP}$. The name 'vapour' is sometimes restricted to the gas phase in contact with its liquid phase (e.g. butane is said to be a gas out of the bottle, but a liquid-vapour mixture inside the bottle).

The phase change from vapour or gas to liquid is sometimes, particularly in cryogenics, named liquefaction.

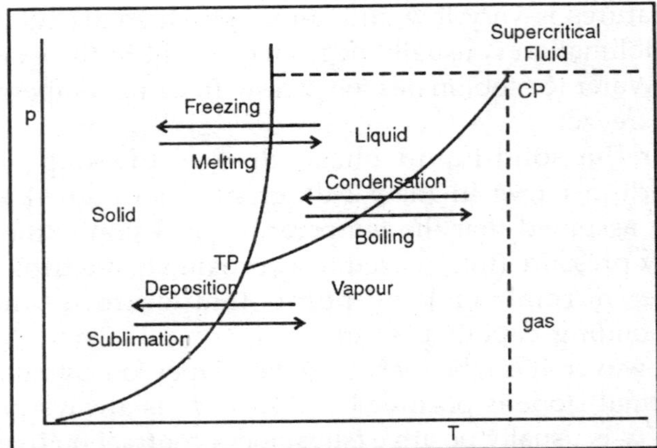

Fig. Nomenclature for Phases and Phase Changes for a Pure Substance, and *p-v-T* Diagram.

The first liquefaction of a gas was made in 1780 by G. Monge, by passing compressed SO_2 through a coil tube immersed in a salt-ice mixture.

PHASE CHANGE VAPOUR FRACTION

In a thermal analysis, when thermodynamic variables are plotted versus temperature at constant pressure, some of them change slope at the phase change (e.g. the Gibbs potential), others suffer a discontinuity in their values (e.g. the intensive volume, being a first derivative of the Gibbs potential), and finally others become ill-defined (e.g. the thermal expansion coefficient, being a second derivative of the Gibbs potential.

The state of equilibrium is just defined by one variable, T or p, but in order to fully define the physical state one needs to know the relative proportions of the phases. Any of the additive variables that have a discontinuity serves this purpose if its averaged value is given. The traditional way to measure the proportions is the vapour mass fraction, x, also known as vapour title or vapour quality:

$$x \equiv \frac{m_V}{m_V + m_L}$$

Some care is needed with this variable since the x's are used for molar fractions (not mass fractions) in all but this

Chapter. Besides, it only refers to one of the phases, the mass fraction of the liquid phase being 1-x.

Any specific variable (i.e. extensive variable by unit of mass) for the whole two-phase system is defined like for the enthalpy:

$$h \equiv \frac{H}{m} = \frac{m_V h_V + m_L h_L}{m_V + m_L} = xh_V + (1-x)h_L = h_L + xh_{LV}$$

i.e. the same applies for s, u, v, g, ϕ etc. Concerning the values of functions in the liquid phase and on the vapour phase (e.g. h_L and h_V) it is imperative to be aware that only one reference state can be chosen for all of them. Thus, if one applies the perfect liquid model to the liquid and the perfect gas model to the vapour, one may write h_L-h_{L0}=$c_L(T-T_0)+(p-p_0)/\rho_L$ and h_V-h_{V0}=$c_p(T-T_0)$, but both h_{L0} and h_{V0} cannot be chosen as zero because h_{V0}-h_{L0}=h_{LV0}. The reference state is a free choice to the user, h=0 and s=0 at the saturated liquid state at the triple point being a natural generic choice, although there are other choices in the literature: ASHRAE sets h=0 and s=0 for saturated liquid at T=-40 °C (=-40 °F); IIR sets h=200 kJ/kg and s=1 J/(kgK) for saturated liquid at T=0 °C; others set h=0 and s=0 for saturated liquid at its normal boiling point.

Notice that the thermal capacity of a biphasic single-component system is infinite since $c_p=\partial h/\partial T|_p$ and at p=constant the system increases h with dT=0.

Once a reference is taken, e.g. the liquid state at the triple point; the value for a vapour state may be composed in the three different manners sketched in Fig. that with the perfect-substance models give, for the enthalpy:

1) h_V-h_0=h_{LV0}+$c_p(T-T_0)$,
2) h_V-h_0=$c_L(T_1-T_0)$+h_{LV1}+$c_p(T-T_1)$
3) h_V-h_0=$c_L(T-T_0)$+h_{LV}.

Any path should give the same result (under the accuracy of the model) since they are state functions, but the first is shorter if the vapourisation enthalpy at the triple point is known (e.g. h_{LV0}=2.50 MJ/kg for water) and the second only needs the vapourisation enthalpy at the boiling point that is ready available (e.g.h_{LV0}=2.26 MJ/kg for water).

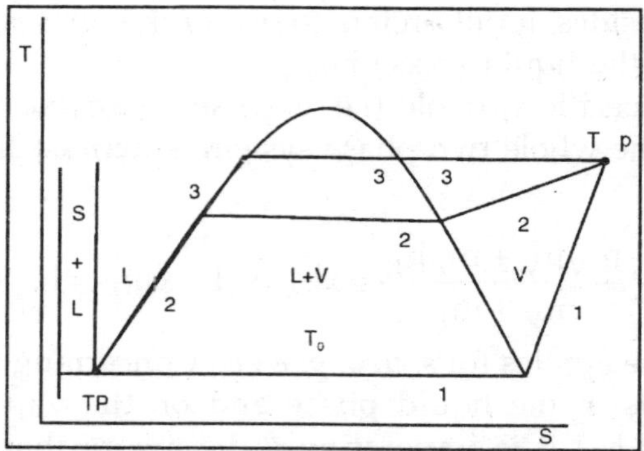

Fig. Usual Reference State for Energies and Entropies

Notice also that mean variables do not measure any local state but the weighted average according to the mass proportions of the phases, as seen in When two-phase states of a system are represented in a thermodynamic diagram, the point corresponding to those mean variables should be interpreted as auxiliary points that show the proportion of the two phases, whose properties, constant along an isobar (or isotherm), are those at the extremes of the isobar, as shown in Figure. The linear application between an specific variable like h in (6.6) and x is known as the lever rule (i.e. the amount of each phase is proportional to the distance to the other extreme).

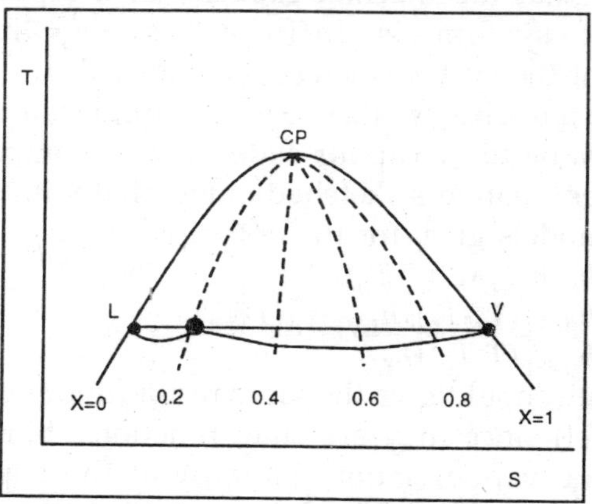

Fig. A Biphasic State B Corresponds to the Averaged Properties (in this Case Specific Entropy) of the Coexisting Phases: Liquid, L, and Vapour, V.

The two bounds to the two-phase region, are known as the saturated liquid curve (x=0) and the saturated vapour curve (x=1).

Mixed states inside the two-phase region, below the x=1 curve, are known sometimes as wet vapour states, and those beyond x=1 superheated vapour states.

The vapourisation enthalpy monotonically decreases from the triple point to the critical point where it cancels, what can easily be appreciated in a p-h diagram on the width of the two-phase region (e.g. for water, h_{LV}=2500 kJ/kg at 273 K, h_{LV}=2260 kJ/kg at 373 K and h_{LV}=0 at 647 K).

Phase change problems occur in many natural and artificial processes: the hydrological cycle, casting, welding, soldering, boilers, freezers, cooling towers, etc.

Exercise

A type of commercial butane bottle has a capacity of 26.1 litres and holds 12.5 kg of butane (in practice it is a hydrocarbon mixture as obtained from crude-oil distillation, not a pure component).

Evaluate:

- If it is possible to have this amount of butane inside this volume.
- The pressure inside for the three cases: cold operation (10°C), nominal operation (20°C) and maximum allowed (50°C).
- The volume of liquid and vapour.
  =Una botella de butano comercial de las antiguas tiene una capacidad de 26,1 litros y contiene 12,5 kg de butano (supóngase puro, aunque en realidad es una mezcla obtenida en el destilado de petróleos). Se pide:
- Determinar si es posible que haya esa cantidad de butano en ese volumen.
- Presión interior en los tres casos siguientes: operación en frío a 10°C, operación normal a 20°C y maxi· permitido a 50°C.
- Volumen de líquido y de vapour.

Solution

- If it is possible to have this amount of butane inside this volume.
  The density of liquid n-butane is found from Thermal Data tables to be 584 kg/m^3, so that up to 584'"0.0261=15.2 kg might be put inside this volume; however, no bottle should be fully filled because of a possible burst by thermal expansion of the liquid.
- The pressure inside for the three cases: cold operation (10°C), nominal operation (20°C) and maximum allowed (50°C).
  Using Antoine equation:

$$\text{In}\left(\frac{p}{p_0}\right) = A - \frac{B}{\frac{T}{T_0} + C}$$

with A=13.98, B=2292 and C=?27.86 for T_0=1 K and p_0= 1 kPa, one gets p_v(10 °C)=150 kPa, p_v(20 °C)=210 kPa and p_v(50 °C)=500 kPa. Notice that butane would not go out of the bottle for temperatures below its normal boiling point (0 °C for n-butane).

The Volume of Liquid and Vapour

The liquid level varies a little with temperature. Let us find its value at 50 °C. One may establish a set of equations $m=m_L+m_V$=12.5 kg, $V=V_L+V_V$=0.0261 m^3, $m_L/V_L=r_L$=584 kg/m^3 (with the perfect liquid model), and $m_V/V_V=r_V=p/(RT)$= 150 × 10^3 /((8.3/0.058)283)=3.7 kg/m^3 with the perfect gas model. These are 4 equations with 4 unknowns (m_L,m_V,V_L,V_V), but many times it is simpler and accurate enough to think that all the mass is liquid and thus it occupies $V_L=m/r_L$=12.5/584=0.022 m^3 and the rest (26-22=4 litres) is vapour with a mass $m_V=r_VV_V=pV_V/(RT)$=0.046 kg.at 50 °C (the vapour mass would be 0.015 kg at 10 °C, and 0.021 kg at 20 °C, with this ideal gas model).

The perfect substance models just used may be good enough in many cases, but a better model might be necessary, at least to judge the degree of approximation. The corresponding state model gives for the vapour at 50 °C $m_V=r_V V_V=pV_V/(ZRT)=0.052$ kg, where Z(pR·TR)=0.87 has been found using Z-p_R diagram (in Thermal Data) with pR=p_{50}/p_{cr}=0.5/3.8=0.13 and TR=T_{50}/T_{cr}=323/425=0.76. For the liquid density, the dilatable liquid model yields=[1(TT_0)] =584[11.810^{-3}(5015)]=547 kg/m^3. The best available data (NIST) gives, for T=323 K, p_v=494.8 kPa =542.3 kg/m^3 and 12.22 kg/m^3, i.e. 23.0 litres of liquid and 3,1 litres of vapour.

Comments

Typical commercial butane in Spain has 56% n-butane, 25% propane, 17% iso-butane, 2% pentane, 0,1 g/kg H_2O and 1 mg/kg mercaptans, with T_b=0.5 °C=580 kg/m^3 at 20 °C and p_v(50 °C)=520 kPa.

The traditional butane bottle (Fig.) is made from 3.2±0.2 mm steel plate (two halves welded in the middle, with a welded collar and screwed valve on the top, and a welded standing ring at the bottom). All bottles are check for resistance with water at 3 MPa, and for air-tightness at 0.8 MPa, and a sample from the lot suffers a hydraulic end test to check that it breaks at >8.5 MPa.

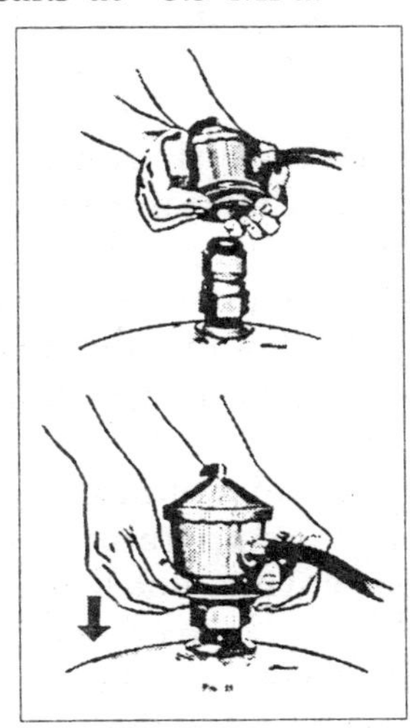

Fig. Domestic Butane Bottle UD-125, with Pressure Regulator K30 (30 mbar).

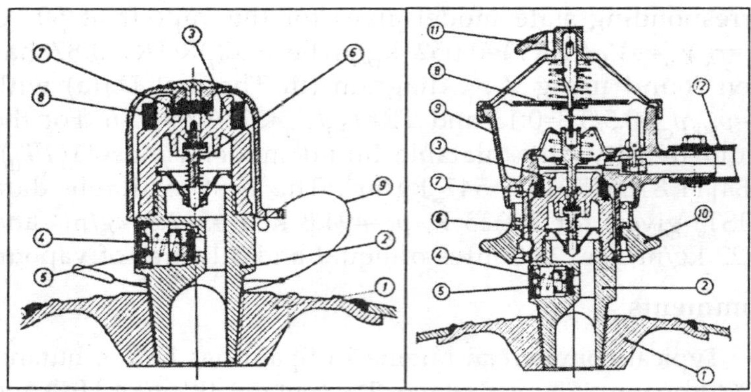

Fig. Details of the Collar, Valve and Regulator (K30) in a Domestic Butane Bottle.

CLOSED SYSTEM EVOLUTIONS

As a closed system, i.e. keeping invariant the total mass, a two-phase system may be processed at constant volume (by heating or cooling), at constant pressure (and thus temperature; e.g. heating or cooling it in a cylinder-piston arrangement), adiabatically (by a quick compression or expansion), or in any more general way.

Notice that heating a liquid inside a fully-filled rigid container does not gives way to a boiling process (except in the rare case of a thermal expansion anomaly); it just increases the pressure up to the mechanical or thermal structural limit of the container, without change of phase; similar to when heating a gas at constant volume. Cooling at constant volume eventually gives way to a liquid/vapour transition in a fully-filled liquid container, and to a vapour/liquid transition in a gas container.

At constant mass and volume, a two-phase system increase or decrease its temperature and pressure when heated or cooled, respectively, but it is no so evident what happens to the liquid level, since on heating, for instance, the liquid expands and the level rises at the same time that some liquid is vapourised (to increase the pressure) and the level falls; which effect is preponderant depends on the value of the thermal expansion coefficient.

Exercise Level Change with Temperature

Statement

A liquid-level indicator, in a two-phase R-12 vessel of 10 cm in diametre and 50 cm height, reads 20 cm from the bottom at 15 °C.

Evaluate:

- Mass of R-12 and initial state.
- Temperature variation that would give way to a rise of 1 mm in the liquid-level indicator.

El indicador de nivel de un depósito de R 12 de 10 cm de diαmetro y 50 cm de altura, un capilar de vidrio conectado en ambos extremos (inferior y superior), marca una altura de líquido de 20 cm en un ambiente a 15 °C.

Se pide:

- Masa de R 12 encerrado y su estado inicial.
- Variación de temperatura que causaría una subida del menisco de 1 mm.

Solution

- Mass of R-12 and initial state.

Refrigerant fluid R-12 is the pure chemical compound dichloro-difluoro-methane, CCl_2F_2. The total mass of fluid in the liquid and vapour phases is computed from density values as: $m=m_L+m_V=r_LAz+r_VA(L-z)$. The cross-section is A=0.0079 m^2, and L=0.5 m. Densities are dependent on thermodynamic state; for the liquid, density can be assumed constant (perfect liquid model) since it varies little with temperature and less with pressure, but for the vapour phase, pressure has a controlling influence. From liquid data tables in Thermal Data, r_L=1330 kg/m^3, and from Antoine's equation in Thermal Data:

$$\ln\left(\frac{p}{p_0}\right) = A - \frac{B}{\frac{T}{T_0} + C}$$

with A=13.79, B=1971 and C=28.30 for T_0=1 K and p_0=1 kPa, one gets p_v(15 °C)=493 kPa.

The ideal gas model gives $\rho_V=p/(RT)=4930000.121/(8.3.288)=24.9$ kg/m^3, whereas the best available data gives $\rho_V=23.65$ kg/m^3.

Thence, the total mass is $m=\rho_L Az+\rho_V A(L-z)=13300.00790.2+24.90.0079(0.5-0.2)=2.08+0.77=2.85$ kg, and the vapour mass fraction $x=m_V/m=0.77/2.85=0.027$. Notice that the simplification of computing just the liquid mass is good enough in most instances, since the uncertainty in liquid density usually outweighs the vapour contribution.

- Temperature variation that would give way to a rise of 1 mm in the liquid-level indicator.
- Differentiating $m=r_L Az+r_V A(L-z)$ we get,

$$0=\frac{d\rho_L}{dT}Az+\rho_L A\frac{dz}{dT}+\frac{d\rho_V}{dT}A(L-z)-\rho_V A\frac{dz}{dT}$$

and, solving for the unknown, in short notation (subindex T meaning derivative):

$$z_T=-\frac{\rho_{L_T}Z+\rho_{V_T}(L-z)}{\rho_L-\rho_V}$$

If we evaluate this expression using the perfect fluid models $r_{L,T}=0$ (incompressible liquid) and $\rho_{V,T}=p_T/(RT)-p/(RT^2)=0.74$ kg/(m^3×K), where the derivative of the vapour pressure with temperature, p_T, is obtained explicitly from Antoine equation, particularising at the working point, resulting finally in $z_T=-0.0002$ m/K, i.e. the level would decrease 0.2 mm per kelvin of temperature rise, due to the liquid loss by vapourisation.

However, if a more accurate model is used (e.g. a linear expansion model for the liquid, with $a=0{,}0025$ 1/K, or more detailed tabulated data), the results are quite different: $\rho_{L,T}=-3.4$ kg/(m^3×K), $\rho_{V,T}=0.81$ kg/(m^3×K), and $z_T=0.0003$ m/K, i.e. the level would increase 0.3 mm per kelvin of temperature rise, because the effect of the liquid expansion outweighs the effect of the vapourisation.

Comments

Chlorofluorocarbons (CFC, like R-12), and hydrochlorofluorocarbons (HCFC, like R-22), are being replaced by hydrofluorocarbons (HFC, like R-134a) and their mixtures, due to their environmental impact on the depletion of the ozone layer and, to a lesser extent, on global warming.

When a two-phase system is suddenly expanded, the liquid phase starts immediately to boil-off, and the vapour phase usually condenses a little (heavy-molecule substances may expand along the gas phase), both immediately reaching the equilibrium temperature at the new pressure, so that, neglecting the entropy of dispersion of one phase in the other as is always done, the process may be considered isentropic.

However, when a two-phase system is suddenly compressed, the liquid phase immediately adapts to the new pressure without any temperature variation, whereas the vapour phase gets compressed adiabatically with a large temperature increase, so that afterwards there is an entropy increase by heat exchange between both phases.

Open System Evolutions

The most frequent control volume systems operate in a one-inlet one-exit steady configuration; two-phase systems may be biphasic at the entrance, at the exit or at both sides. Typical problems are the throttling of a saturated liquid, the expansion of a vapour in a turbine, the mixing of some liquid and vapour streams of the same substance, etc.

Exercise Peabody Throttling

Statement: Vapour mass-fraction in a two-phase system may be evaluated by a small sample throttling (sometimes called Peabody calorimetre).

Deduce:

- Relation between vapour mass-fraction and temperatures and pressures before and after the throttling.
- Minimum vapour mass-fraction detectable and conditions in that case.

Deuna corriente estacionaria de vapour húmedo se extrae un pequeño gasto que se expande hasta la presión ambiente en una vαlvula, con el fin de calcular la fracción mαsica de vapour (calidad) de la corriente (suele llamarse calorímetro de Peabody). Se pide:

- Relación entre la calidad y las temperaturas y presiones antes y después de la valvula.
- Calidad mínima detectable y condiciones necesarias de la corriente.

Solution

- Relation between vapour mass-fraction and temperatures and pressures before and after the throttling.

The process described is shown in Fig.

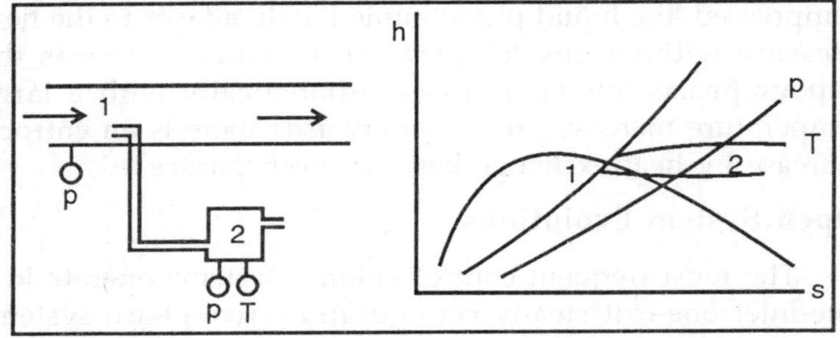

Fig. Peabody Throttling

At point 1, the internal conditions to be measured, T and p are related, thus there is only one data (the other being redundant), say p_1 because it is more accurate (although more expensive to measure that T_1). For the same reason, point 2 must lie outside the two-phase region to have two data: p_2 and T_2. Thus, if p_2 is fixed, e.g. atmospheric pressure, p_2=100 kPa, then T_2 must be >100 °C.

With these measurements, the energy balance, $h_1=h_2=h_{1L}+xh_{1LV}$, helps to compute the vapour mass-fraction x once the enthalpy values are found (by means of the Mollier diagram or data tables). The model of perfect liquids and gases only gives a good approximation at very low pressures, yielding $x=(h_2-h_{1L})/h_{1LV}=1-c_{pv}(T_1-T_2)/[h_{LV0}-(c_L-c_{pv})(T_1-T_0)]$.

- Minimum vapour mass-fraction detectable and conditions in that case.

For a fixed p_2, e.g. atmospheric pressure, p_2=100 kPa, looking at Mollier diagram one finds that to have the largest span of x one needs the minimum T_2, T_2=100 °C, and even so the minimum x is bound to x=0.926 (from the Mollier diagram), corresponding to p_1=4 MPa and T_1=250 °C.

Notice that the model of perfect liquids and gases gives a monotonic decrease of x with T.

In a flash chamber, non-saturated liquid is throttled into a chamber kept at lower pressure by an ejector or vacuum pump.

A portion of the liquid flashes into vapour that, in the gravity field, settles upwards (a demister may help to avoid drops carry over), and can be extracted separately (flash separators) or locally condensed on cold coils and the drops collected on a plate and extracted with a liquid pump (flash distillators).

The latter is used a lot to purify the solvent in a liquid mixture, as for desalting seawater (liquid seawater enters a flash chamber, and two other liquid streams are extracted: brine and distillate).

Some unsteady control volumes are also of interest to phase-change processes, as when mass (liquid or vapour) is added or extracted from a two-phase reservoir, or when steam is used to heat up some water mass (a typical example being milk steaming in coffee machines).

Two-phase bottled compressed gases are routinely used as a source of gas (e.g. camping gas), or as a source of liquid spray (e.g. insecticides). A quasi-steady process develops when a small constant mass flow rate of vapour goes out of a two-phase bottle; its energy balance is $d(m_L e_L + m_V e_V)/dt = -h_V$, and, using the perfect substance models one finally gets a relation between the necessary heat input and the mass flow rate:

$$Q = mh_{LV}\left(1 + \frac{v_L}{v_{LV}}\right) \approx mh_{LV}$$

The temperature (and consequently the pressure) in the interior decreases (due to the vapourisation energy) until at steady state this heat flux is furnished by heat transfer from the environment across the wall of the bottle, at the same rate.

Exercise Butane Supply from a Bottle

Statement: A mass flow rate of 0.6 kg/h of butane is to be supplied from a bottle of 30 litres holding 15 kg of butane. In order to keep the internal pressure at 400 kPa, the bottle is put inside a water bath. Evaluate:

- Bath temperature needed, and initial volume and mass of the gas inside.
- Heat exchanged and time of use.

Se quiere obtener un gasto de 0,6 kg/h de gas butano disponiendo de una botella de 30 litros conteniendo inicialmente 15 kg de butano. Para ello se va a sumergir la botella en un baño con objeto de mantener la presión interior en un valor constante de 400 kPa.

- Temperatura del baño, volumen y masa inicial de gas.
- Calor intercambiado. y tiempo de funcionamiento.

Solution:

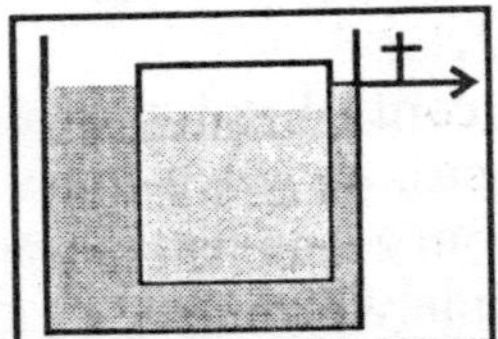

Fig. Sketch of the Bottle in the Bath.

- Bath temperature needed, and initial volume and mass of the gas inside.

Using Antoine equation:

$$\text{In}\left(\frac{p}{p_0}\right) = A - \frac{B}{\dfrac{T}{T_0} + C}$$

with A=13.98, B=2292 and C=27.86 for T_0=1 K and p_0= 1 kPa, the two-phase temperature for 400 kPa is 315 K (42 °C). The liquid volume may be estimated as V_L=m/=15/573=0.026 m^3, and thus 30-26=4 litres of vapour, with a mass of (ideal gas model) m_v=$pV/(RT)$=$40010^3 410^{-3}$/((8.3/0.058) 315)=0.034 kg (a vapour mass fraction of x=0.0023).

- Heat Exchanged and Time of Use.

The energy balance, dE=dW+dQ+h_Vdm, may be reduced to =0.0001736510³=60 W. The autonomy of the bottle is just =25 hours. When vapour is let out of a two-phase vessel, it is usually throttled (without doing work) to the supply pressure, always in the gas phase and with negligible temperature change. If the vapour were expanded doing external work, as in a turbine, it might get fogged (enter the two-phase region again). However, in the case of the liquid phase being let out of a vessel (through a valve in the bottom), the liquid will get bubbly even if just throttled, decreasing its temperature according to the decrease in pressure, what may be used for refrigeration.

Exercise Self-refrigeration in a Propane Vessel

Statement: Internal pressure in a propane vessel of 100 m^3 is kept at 0.6 MPa by bleeding a small amount of liquid, throttling it inside the vessel to atmospheric pressure, passing it through a heat exchanger, and finally venting it out. For atmospheric conditions at 30 °C and 100 kPa:

- Make a sketch of the equipment and the process in the *T-s* diagram.
- Inside temperature.
- Vapour mass-fraction just after throttling.
- Enthalpy changes at each stage.
- Relation between bled mass flow-rate and overall thermal conductance of the vessel.

Para mantener una presión de 0,6 MPa en un tanque de propano de 100 m3 se toma un pequeño flujo de propano del interior, se expande en una vαlvula hasta la presión atmosférica, se pasa por un cambiador de calor (todo ello dentro del tanque) y finalmente se expulsa al exterior, que estα a 30 °C y 100 kPa. Se pide:

- Esquema de la instalación y representación del proceso en un diagrama *T-s*.
- Temperatura dentro del tanque.
- Fracción mαsica de vapour después de la vαlvula.
- Variación de entalpía del propano desde el tanque al exterior.

- Relación entre el gasto de propano y la transmitancia térmica del tanque.

Solution:

- Make a sketch of the equipment and the process in the *T-s* diagram.

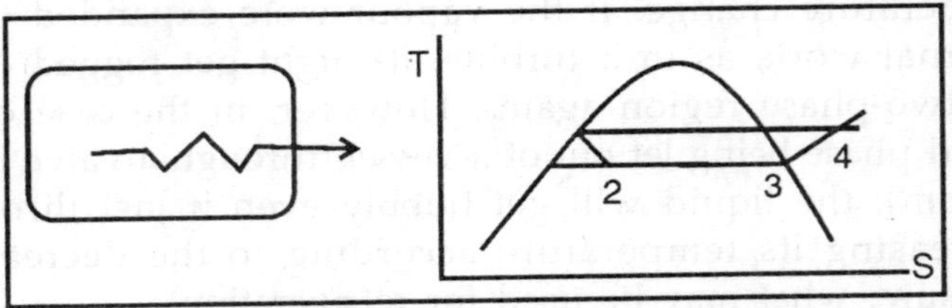

Fig. Sketch of the Equipment and the Process

- Inside temperature.

Using Antoine equation:

$$\ln\left(\frac{p}{p_0}\right) = A - \frac{B}{\frac{T}{T_0}+C}$$

with A=13.71, B=1873 and C=25.10 for T_0=1 K and p_0= 1 kPa, one gets the two-phase temperature for 0.6 MPa, which is 281 K (8 °C); point 1.

- Vapour mass-fraction just after throttling.

After throttling, Antoine equation gives 231 K (-42 °C) for the two-phase temperature at atmospheric pressure (its boiling point). The energy balance from 1 to 2 is $h_1=h_2=h_{2L}+xh_{2LV}$, and, using perfect substance models one gets $x=(h_1-h_{2L})/h_{2LV}=c_L(T_1-T_2)/h_{2LV}$=2120(281-241)/430000=0.25.

- Enthalpy changes at each stage.

$h_2-h_1=0$, $h_3-h_2=(1-x)h_{LV}$=320 kJ/kg and $h_4-h_3=c_pT$=78 kJ/kg, but it is safer to neglect the last term, i.e. the possible refrigeration on the gas phase, to reduce the cost of the heat exchanger.

- Relation between bled mass flow-rate and overall thermal conductance of the vessel.

The energy balance $\Delta E=Q+W+\Sigma h_{te}dm_e$ with one opening and at quasi-steady state, becomes:

$$Q = m\Delta h = KA\Delta T$$

meaning that, for a given global heat conductance with the ambient, *K*, and given contact area *A*, the boil-off flow-rate is proportional to the temperature jump wanted (dictated by the maximum pressure jump allowed).

METASTABLE STATES

Metastable states are non-equilibrium states that appear stable under some circumstances, because the ambient noise has not enough energy or time to force a transition, but that under a sizeable perturbation, they evolve to a widely different equilibrium state (e.g. think on the effect of a <1 mJ spark on a H_2/O_2 mixture, or the trigger in a hot or cold pad).

In order to trigger the phase change, some jump in chemical potential must be applied, like the activation energy in combustion processes.

To be more precise, consider phase changes from the liquid state at constant pressure. Liquids may be superheated above their boiling point without changing to vapour (as often happens in microwave ovens), and undercooled below their freezing temperature (as easily done with distilled water in a test tube).

Notice that we are using the words undercooling, subcooling, and supercooling, as synonyms.

Vapours may also be undercooled below their condensation temperature, but solids are very difficult to superheat over their melting temperature.

Similarly, at constant temperature, the pressure in a liquid may be decreased below its equilibrium bubbling pressure, even to negative pressures as in tall trees (in 1850, Berthelot subjected purified water to 5 MPa of tension), along metastable states, LL′ in Figure and vapours compressed above their condensing pressure along metastable states VV′ in Figure.

However, states along L′V′ are always mechanically unstable (they would had $\hat{e}$<0) and cannot be got in practice. The loci of L′ and V′ for different isotherms is known as the spinodal curves (spinode, Gr. cusp), liquid spinodal for L′ and vapour spinodal for V′.

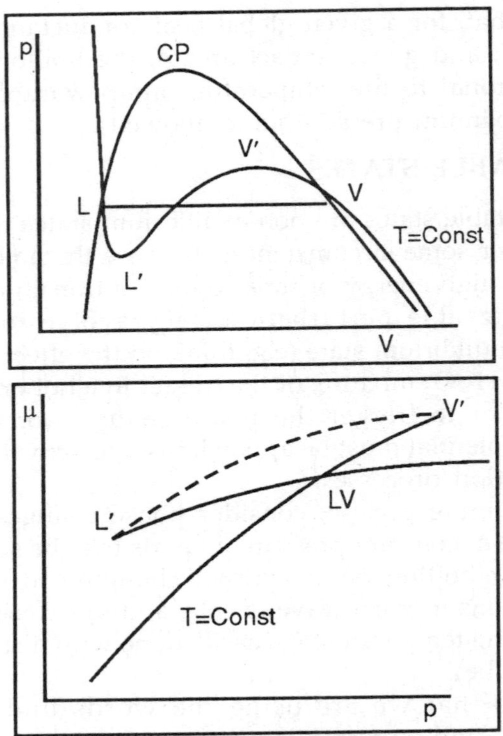

Fig. An Analytical Isotherm T=const.(e.g. a van der Waals one) and the Real discontinuities: a Jump in the Specific Volume (straight line LV in the p-$v$diagram), and a change of slope in specific Gibbs potential (chemical potential, ì).

Vapour states from V to V', and liquid states from L to L', called metastable states, can be obtained under controlled conditions of small perturbations in very pure substances.

The ì-p diagram in Fig. that may be built from the p-v diagram by means of the Gibbs-Duhem equation, $0=-sdT+vdp+d\mu$ ($\mu=\mu_0-v dp$ along an isotherm), also gives an indication of the extra energy metastable states need to overpass the barrier of unstable states (dashed line). By the way, it is also seen that, for an analytical isotherm, the points corresponding to the phase change are those of equal chemical potential, those that verify $vdp=0$, or $(p-p_{LV})dv=0$, i.e. when the two areas between the curve and the straight line in the *p*-*v* diagram of Fig. are of equal area, what is known as Maxwell's rule.

Quick processes are prone to develop metastable states, as when a vapour expands in a nozzle, where it may go without phase change down to pressures below the transition one and suddenly condense in an abrupt isenthalpic jump (generally associated with the chock wave in the diverging side of a supersonic nozzle. Conversely, metastable states are prone to develop fast processes; consider the difficulty in freezing a pan with water in a fridge (ice starts at the boundaries, and take a lot of time); if, however, the liquid were undercooled to say 5 °C (very difficult with tap water, but not so with distilled water, which may be cooled down to 42 °C under extreme care), upon a small perturbation, a sudden freezing takes place (with a temperature increase up to 0 °C due to the release of the fusion enthalpy); you can compute the amount of ice formed, by an energy balance. Flash vapourization is advantageously used in superheated fuel injectors too.

Metastable states may classified in terms of the equilibrium line done through, and the direction:

- Through solid-liquid equilibrium line
  - Superheated solid. It is difficult to keep a substance in the solid state when heated over its melting point; that is why melting-point measurements are more accurate than freezing point ones.
  - Subcooled liquid. As said above, it is not difficult to cooled a liquid below its freezing temperature without it changing to the solid state. There is a lot of supercooled water in small droplets on middle and high clouds in the atmosphere (posing a risk to aviation because the metastable liquid droplets freeze on contact with the plane and spoil aerodynamic lift).
- Through liquid-vapour equilibrium line
  - Superheated liquid. As said above, it is not

difficult to get superheated liquid water (at >100 °C) in a microwave oven, and see it 'explode' by sudden boiling when a tea-bag or sugar is added. A sudden pressure loss in a liquid container or piping usually yields metastable liquid states too.

- Subcooled vapour. A classical example is the Wilson cloud chamber used for detecting particles of ionizing radiation.

- Through solid-vapour equilibrium line. These changes rarely yield metastable states.

The degree of a metastable state (e.g. degree of superheating, or degree of under cooling) can be measured as the temperature departure from the equilibrium line at constant pressure, or as the pressure departure from the equilibrium line at constant temperature.

Material properties of metastable phases (e.g. density, thermal capacity...) are simple extrapolation of their stable-state properties (e.g. thermal expansion of supercooled water (also named sub-cooled water) is negative with positive slope of $\delta\alpha/T$, with $\alpha=-1.5\cdot10^{-3}$ 1/K at 30 °C and $\alpha=1.2\cdot10^{-3}$ 1/K at 150 °Cand 100 kPa). However, one must be careful with phase-change relations like $\Delta s=\Delta h/T$, which only hold for equilibrium transitions; for metastables transitions, the transition enthalpy must be supplemented with the difference in thermal capacities from the stable and the metastable phases times the undercooling or overheating.

It seems that the first metastable states were recognised by Farenheit in 1714 when studying the freezing of water (he reported different undercooling degrees, depending on the cooling rate). The first theory of metastable states is due to Gibbs (1878).

Creation of a new phase requires a new local arrangement of matter (the nucleation process) which demands some activation energy that builds up on metastable states. Fluid-to-fluid phase-transitions in pure substances are the vapourization of a liquid (boiling) and the condensation of a vapour (e.g. in steam condensers). When a solid phase is involved, the direction of change cause great differences between the melting or vapourization of a solid, in one hand, and the solidification of a liquid or deposition of a vapour, on the other hand, particularly when the nascent solid is in crystalline form instead of amorphous. When mixtures are considered, the phase changes and their nucleation processes become more complicated (e.g. condensation from supersaturated solutions, constitutional supercooling), and new liquid-to-liquid phase changes might appear. The formation of solid phases is studied in the science of crystal growth, so important in materials science and in natural frosting (including the common case in cloud formation).

VAPOURIZATION KINETICS

The growth of a different phase at the expense of another phase (as the growth of bubbles from a boiling liquid, or the growth of droplets from a condensing vapour), usually takes place at pre-existing nucleation sites (either in the heterogeneities of the container walls, or at the heterogeneities of suspended particles within the fluid), because the nucleation process in a clean fluid within smooth walls is most difficult. In the latter case, named homogeneous nucleation, the nuclei are created by large equilibrium fluctuations, which are proportional to fluid compressibility. Small nuclei, however, are unstable if their size is below a critical radius r_{cr}=2óΔp (ó being the liquid-vapour interface tension, and Δp the underpressure), so that a large fluctuation driven by a high superheating or undercooling is required for homogeneous nucleation.

Consider the vapourization process taking place when a liquid in a reservoir is depressurised below its vapour pressure at the initial temperature; this is the best way to avoid temperature gradients and a quick phase transition (if heating were done, gradients would be proportional to heating speed). In any case, there is a correspondence between superheating and depressurization, $\Delta p/\Delta T=(h_{lv}/(Tv_{lv}))$, dictated by Clapeyron's equation.

The depressurization process is isenthalpic because it is adiabatic (sudden), and without work exchange with the surroundings. The initial liquid state may be a compressed-liquid state (as in Fig.), a saturated liquid (state 's' in the saturated liquid line), or a (metastable) superheated liquid state. The final state may be point 2 in Fig. in the normal case of partial vapourization (the two phases may later segregate under gravity), or total vapourization at point 3 for deep-enough depressurizations. The required vapourization enthalpy is obtained from the internal energy of the superheated liquid, leaving a colder two-phase state.

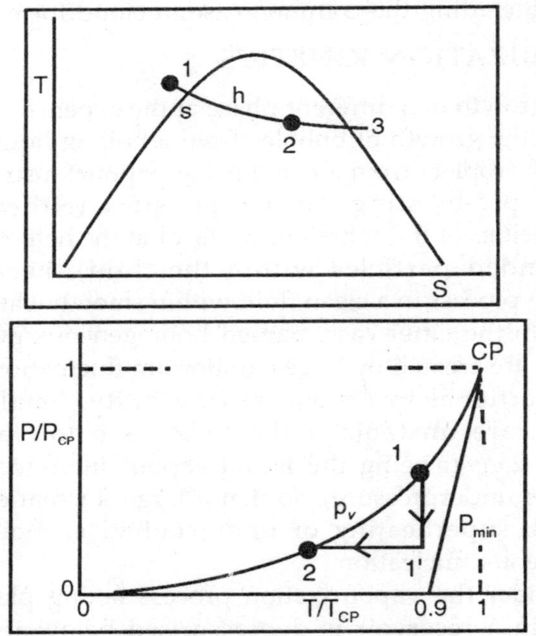

Fig. Liquid flashing by sudden depressurization: a) isenthalpic evolution in the *T-s* diagram; b) metastable evolution in the phase diagram:

Many experiments have been carried out with a pressurized liquid in a vertical test tube (from 1 mm to 300 mm in diametre) that is suddenly opened to vacuum (or to ambient pressure, depending on liquid volatility). If the liquid has been degasified, the rarefaction pressure wave reaches the liquid bottom in less than 1 ms, but homogeneous nucleation cannot take place except close to the critical point temperatures (when the spinodal line (p_{min} in Fig.) would be crossed, and boiling can only starts at the free surface (providing the glass wall is clean enough), with a fuzzy bubbling that ejects droplets and produces a two-phase downstream flow.

Within about 10 ms, a vapourization front develops and attains a steady propagation downward the liquid, with a speed of order 1 m/s (0.1..10 m/s) depending on the liquid substance and the depressurization (or superheat), but independent of the tube diametre. The upstream liquid remains quiescent and at the initial temperature (except for a cooled boundary layer close to the advancing front), in a metastable equilibrium. The front is more or less planar but clearly unsteady and random in detail, with several millimetres in thickness; the typical droplet size is about 1 mm. The developed two-phase flow consists of vapour as the continuous phase and small liquid droplets as the dispersed phase, 10..100 ìm in size, some times in chain-like streaks. Exit pressure is dictated by downstream details (usually a hole with chocked flow).

This travelling vapourization wave is very similar to the laminar deflagration wave in premixed combustion, with similar speeds, proportional to the square root of thermal diffusivity in both cases, etc. The depressurization process just described has been applied to many engineering problems, like the safety analysis of loss of coolant accidents in pressurised water nuclear reactors, and some natural processes like volcano eruptions (when hot lava, under pressure by overlying rock, is released).

TYPE OF PROBLEMS

Besides housekeeping problems of how to deduce one particular equation from others, the types of problems in this chapter are:

- Find the new state and the energy variation when a two-phase reservoir is heated or cooled.

- Find the mass flow rate escaping from a two-phase reservoir when the heat input is prescribed or viceversa (this is of interest in boilers).
- Find the new state and when a steady flow of liquid is depressurised.
- Find the new state and the work output when a steady flow of vapour is expanded and some condensate appears.

GENERATION OF STEAM

STEAM

Water can exist in the form of solid, liquid and gas as ice, water and steam respectively. If heat energy is added to water, its temperature rises until a value is reached at which the water can no longer exit as a liquid. We call this the "saturation" point and with any further addition of energy, some of the water will boil off as steam. This evapouration requires relatively large amounts of energy, and while it is being added, the water and the steam released are both at the same temperature. Equally, if we can encourage the steam to release the energy that was added to evaporation it, then the steam will condense and water at the same temperature will be formed.

Why use Steam

Steam is produced by evaporation of water, which is a relatively cheap and plentiful commodity in most parts of the world. Its temperature can be adjusted very accurately by the control of its pressure, using simple valves; it carries relatively large amounts of energy in a small mass, and when it is encouraged to condensate back to water, high rates of energy flow (into the material being heated) are obtained, so that the heat using plant does not have to be unduly large.

Thus steam is most economical,flexible and versatile tool for industry wherever heating is required.

Liquid Enthalpy

Liquid enthalpy is the "Enthalpy" (heat energy) in the

water when it has been raised to its boiling point to produce steam, and is measured in kJ/kg, its symbol is h_f. (once known as "Sensible Heat")

Enthalpy of Evaporation

The Enthalpy of evaporation is the heat energy to be added to the water (when it has been raised to its boiling point) in order to change it into steam. There is no change in temperature, the steam produced is at the same temperature as the water from which it is produced, but the heat energy added to the water changes its state from water into steam at the same temperature.

When the steam condenses back into water, it gives up its enthalpy of evaporation, which it had acquired on changing from water to steam. The enthalpy of evaporation is measured in kJ/kg its symbol is h_{fg}. Enthalpy of evaporation is also known as latent heat.

The temperature at which water boils increases as the pressure increases. From this it is evident that as the steam pressure increases, the usable heat energy in the steam (enthalpy of evaporation) which is given up when the steam condenses, actually decreases. The sum of the two enthalpies is known as the enthalpy of saturated steam. This enthalpy is the total heat energy, which is stored in the steam.

Table. Extract from the Steam

Pressure(Bar)	Temperature °C	Enthalpy in kJ/kg			Volume (m^3/kg)
		Water (h_f)	Evaporation (h_{fg})	Steam(h_g)	
0	100	419	2257	2676	1.673
1	120	506	2201	2707	0.881
2	134	562	2163	2725	0.603
3	144	605	2133	2738	0.461
4	152	671	2108	2749	0.374
5	159	641	2086	2757	0.315
6	165	697	2066	2763	0.272
7	170	721	2048	2769	0.240

STEAM VOLUME

If 1 kg (mass) of water (which is 1 litre, by volume) is all converted into steam, the result will be exactly 1kg (mass) of steam. However, the volume occupied by a given mass depends upon its pressure. At atmospheric pressure 1kg of steam occupies nearly 1.673 cubic metres(m3). At a pressure of 1bar abs, that same 1 kg of steam will only occupy 0.1943 m3. Thus, steam should always be generated and distributed at rated boiler pressure and used at possible low pressure.The volume of 1kg of steam at any given pressure is termed its Specific Volume (symbol Vg).

Steam Quality

In practice, steam often carries tiny droplets of water with it and cannot be described as dry saturated steam. Nevertheless, we find that it is usually important that the steam used for process or heating is as dry as possible.

Steam quality is described by it's "dryness fraction" - the proportion of completely dry steam present in the steam being considered. The steam becomes "wet" if water droplets in suspension are present in the steam space, carrying no specific enthalpy of evaporation. "Wet steam" has a heat content substantially lower than that of dry saturated steam at the same pressure. The small droplets of water in wet steam have weight but occupy negligible space. The volume of wet steam is, therefore, less than that of dry saturated steam.

Volume of Wet Steam = Volume of Dry Saturated Steam * Dryness Fraction

Dryness fraction of the steam depends upon the steam boiler design and capacity. For example, Coil type low capacity boilers produce about 30 to 50% wet steam which is not desired in steam heating applications.

Superheated Steam

As long as water is present, the temperature of saturated steam will correspond to the figure indicated for that pressure in the Steam Tables. However, if heat transfer continues after all the water has been evaporated, the steam temperature will again rise. The steam is then called "superheated", and this "superheated steam" can be at any temperature above that of saturated steam at the corresponding pressure. Superheated Steam is totally dry and follows the gas laws.

Saturated steam will condense very readily on any surface which is at a lower temperature and gives up the enthalpy of evaporation which, as we have seen, is the greater proportion of its energy content. On the other hand, when superheated steam gives up some of its enthalpy, it does so by virtue of a fall in temperature.

No condensation will occur until the saturation temperature has been reached, and it is found that the rate at which we can get energy to flow from superheated steam is often less than we can achieve with saturated steam. Even though the superheated steam is at a higher temperature, superheated steam because of its properties, is the natural first choice for power steam requirements, whilst saturated steam is ideal for process and heating applications.

Chemical Energy

The chemical energy, which is contained in coal, gas or other boiler fuel, is converted into heat energy when the fuel is burned. That heat energy is transmitted through the wall of the boiler furnace to the water.

The temperature of the water is raised by this addition of heat energy until saturation point is reached – it boils. The heat energy which has been added and which has had the effect of raising the temperature of the water is known as the liquid enthalpy.

At that point of boiling, the water is termed Saturated Water. The water in our boiler is now at saturation (boiling) point at 100 °C. Heat transfer is still taking place between the furnace walls and the water. The additional enthalpy produced by this heat transfer does not increase the temperature of the water. It evaporates the water, which changes its state into steam. The enthalpy that produces this change of state without change of temperature is known as the enthalpy of evaporation.

BOILER

The job of a boiler is to supply good quality dry steam at the correct pressure at the right time.

Boilers and the associated fire equipment should be designed for efficient operation. They should also be properly sized. A boiler, which has to cope with a peak load above its maximum continuous rating, will operate at reduced efficiency. Pressure may drop and the resultant priming and carry-over will mean that the boiler is unable to do its job of providing good quality steam.

If a boiler has to work at a small per centage of its rating, radiation losses become significant and, again there is a drop in overall efficiency. Clearly, it is not easy to match boiler plant to what is normally a variable steam load. Two or more boilers are more flexible than a single unit which explains the common arrangement of a large boiler for the winter load with a smaller boiler for the summer load.

Boiler Losses

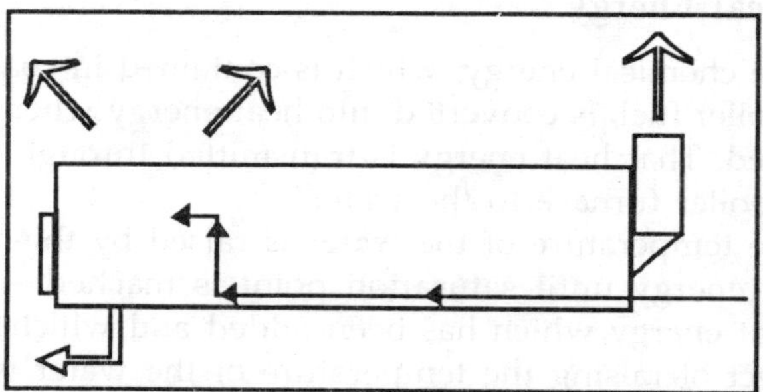

There are four different types of boiler losses:

- Radiation losses
- Flue gas losses (stack losses)
- Blowdown losses
- Losses due to low rate of condensate return from the production plant

The radiation losses are determined through the thermal insulation of the boiler. It should be tried to return as much condensate as possible back to the boiler, so the used heat energy in the boiler can be minimized.

The flue gas and the blowdown losses should be discussed more in detail.

Flue Gas Losses

To achieve a high thermal efficiency, thereby minimizing fuel costs, the amount of combustion air should be limited to that necessary to achieve complete combustion of the fuel.

Excess air is the extra air supplied to the burner beyond the air required for complete or stoichometric combustion. If the supplied air is less, not only will this result in a smoking stack (black smoke indicates the presence of unburned fuel i.e. combustibles in the flue gas), but it will significantly reduce the energy released per unit of fuel. If a burner is operated with a deficiency of air, carbon monoxide and hydrogen will appear in the products of combustion. These combustibles are fuel, and anything in excess of a few hundred parts per million in the flue gas indicates inefficient burner operation.

Too little excess air is inefficient because it permits unburned fuel, in the form of combustibles, to escape up the stack. But too much excess air is also inefficient because it enters the burner at ambient temperature and leaves the stack hot, thus stealing useful heat from the process.

Getting the right mix of fuel and air in your burner is the most critical and difficult step in achieving boiler efficiency. The Oxymiser Combustion Analyser makes sure you always get the right mix of air and fuel.

Blowdown Losses

Boiler blowdown is a very essential function which – if not carried out in the most energy efficient manner – can be an unnecessary source of loss.

Raw water will contain impurities. As the boiler water is evaporated into steam and replaced by make up water, the concentration of the solids in the boiler water will obviously increase. If they are allowed to increase much beyond the recommended level then the functioning of the boiler will be affected. "Foaming" will take place within the boiler resulting in carry-over of water into the steam distribution system, and in the worst situation, malfunctioning of the boiler water level controls can occur.

The original method of "blowing down" a boiler in order to maintain the total dissolved solids (TDS) at an acceptable level was to manually operate a blowdown valve fitted to a discharge pipe at the lowest point of the boiler shell, at the same time hoping that the frequency of operation of the blowdown valve and the length of time for which it is open is sufficient for the needs of the boiler.

Best engineering practice now in the furtherance of energy efficient plant, is to apply the use of an Automatic Blowdown control system (ABCO) which is not necessarily high in capital expenses compared with the payback period in terms of energy saving.

The losses from blowdown may be also be minimized by using the heat in the blowdown water. A certain per centage of the blowdown water will flash off into steam when it enters a region of pressure lower than that which existed within the boiler. This flash steam can be recovered and used by the application of a "Blowdown Heat Recovery Unit" which consists of Blowdown Flash Vessel and Heat Exchanger system to recover the flash steam and use the heat from blowdown water to preheat the Boiler Feed Water. Flash steam is sent to feed water tank for direct heating of water.

The EffiMax 2000 package provides a complete monitoring and data acquisition solution for your boiler performance. EffiMax calculates the efficiency of the boiler based on indirect efficiency computation and computes individually the total amount of losses like stack loss, enthalpy loss, radiation loss and blowdown loss in your boiler. Using the data generated on the system losses, on-line suggestions can then be used to fine tune the system to generate more steam with the lesser quantity of fuel.

Water for the Boiler

The boiler feed tank is the heart of any steam system. It provides a reservoir of returned condensate and fresh makeup water with which the boiler feed pump can replenish the boilers.

The feed tank must be properly sized and allowance made for fluctuations and possible interruptions in supply; it is normal to hold enough water to provide one hour of steam at maximum rating. However, there should be enough free space to cope with the relatively massive return at start-up. Significant quantities of condensate can be lost if this is not provided.

In order to prevent corrosion in boilers and ancillary equipment it is necessary to eliminate both dissolved oxygen and dissolved carbon dioxide from the boiler feed water. Oxygen is the main cause of corrosion and the presence of carbon dioxide in the water presents a pH value of something less than neutral (pH 7) causing the water to be acidic. The ideal pH value for boiler feed water is around pH 9. A higher pH value can cause what is called "caustic embrittlement" in the boiler, particularly a boiler with riveted joints. A Feed Water Tank Systemfrom Spirax provides complete solution to manage the boiler feed water for any capacity boilers.

A Deaerator Head ensures vigorous mixing of steam and feedwater to reduce dissolved oxygen content. This action reduces the need for oxygen scavenging chemicals to a minimum.

EFFICIENCY IN STEAM GENERATION

Every process needs to be optimised for energy efficiency The process of Generating energy is no exception. Efficiency of energy generation is a key factor in the overall profitability of any process plant. Spirax Marshall has a range of unique products, systems and software to help you keep track of exactly how much energy you consume and where it goes.

Our Steam Metreing and Effimax range of products, help you keep an accurate track on parametres that show efficiency of energy generation. Our automatic blowdown control systems ensure a fine control on activities like boiler water blowdown, which is a potential area for energy leakage. Boiler heat recovery units, flash vessels, steam injectors and the

Effipro range of products are some of the solutions designed to recover every bit of energy that might be lost in the boilerhouse itself.

TRIPLE POINT AND CRITICAL POINT

TRIPLE POINT

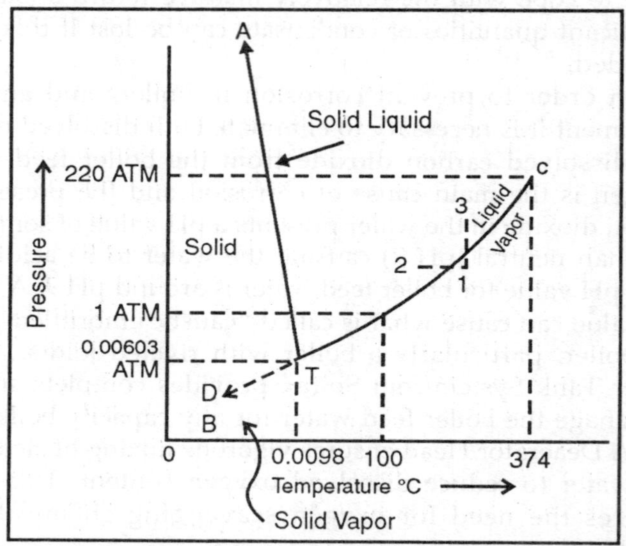

Triple point is the intersection on a phase diagram where three phases coexist in equilibrium. The most important application of triple point is water, where the three-phase equilibrium point consists of ice, liquid, and vapour. Before discussing triple point further, a basic understanding of the lines from Figure, the phase diagram of water, are first considered.

Take the line TC which gives the vapour pressure of liquid water up to the critical point C. Along this line, liquid and vapour coexist in equilibrium. At temperatures higher than that of point C, condensation does not occur at any pressure.

The line TA represents the vapour pressure of solid ice, which is a plot of the temperatures and pressures at which the solid and vapour are in equilibrium. Finally, line TB gives the melting point of ice and liquid water. The plot shows the temperatures and pressures at which ice and liquid water are in equilibrium.

Note: At the dashed line TD, liquid water can be cooled below the freezing point to give supercooled water.)

The preceding paragraphs show that two phases are in equilibrium along the three solid lines. But when these lines intersect at one point C, three phases coexist in equilibrium. This intersection is the triple point, where a substance may simultaneously melt, evaporate, and sublime.

Example Problems involving the Triple Point of Water

Problem 1: Temperature vs. Pressure

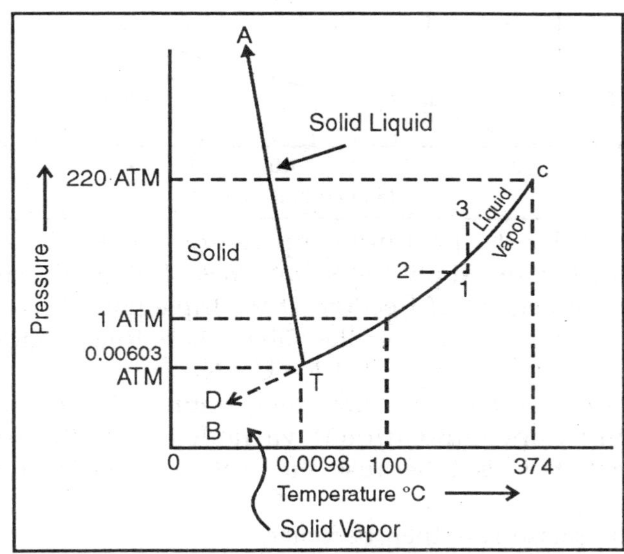

Given the phase diagram for water above, what happens to the melting point as you increase pressure?

The figure shows that as the pressure increases, the melting point increases to a maximum at the triple point. We know the temperature at this point to be zero Celsius, which is the melting point of water.

Describe the changes that occur as a result of moving

across the line from point 1 to point 2. and from point 1 to point 3. In order to get to point 2, the temperature must decrease, while the pressure must increase to reach point 3. However, both process crosses the liquid-vapour equilibrium line in the direction of condensation from vapour to liquid.

Problem 2: Gibbs Phase Rule

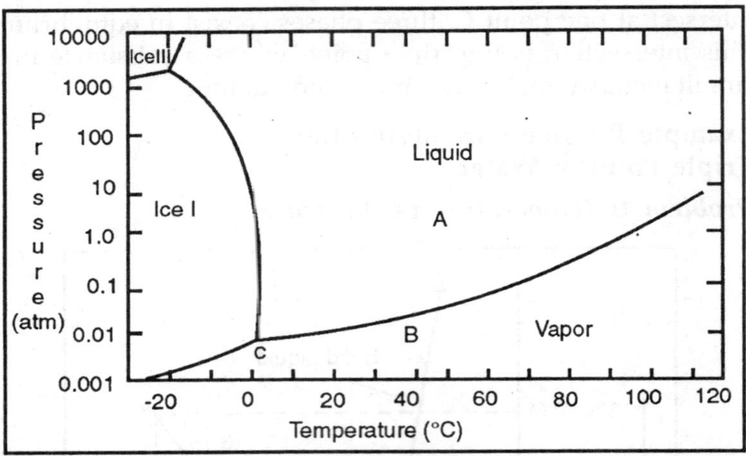

Consider the pressure-temperature phase diagram for water of Figure. Apply the Gibbs Phase Rule to specify the number of degrees of freedom at the triple point C.

This problem calls for the Gibbs Phase Rule, which is

$$P + F = C + N$$

For this system, N=2 since temperature and pressure are the only noncompositional variables. The number of components C is 1 because the system consists of solely water.

The phase rule then becomes,

$$P + F = 1 + 2$$

since we are solving for F, the equation reduces to,

$$F = 3 - P$$

Applying this point to the triple point, the number of phases present P is 3, which makes F = 0.

Now consider the points A and B, find the degrees of freedom, then compare it to point C.

The phase rule F = 3 - P remains the same since it is the same system. At point A, only a single phase is present, so P=1. The number of degrees of freedom, F=2. At point B, which is the boundary between liquid and vapour phases, two phases are in equilibrium, making F = 1.

From these results we can conclude that at the triple point, where F = 0, we have no choice in the selection of externally controllable variables in order to define the system.

CRITICAL POINTS

We will discuss the occurrence of local maxima and local minima of a function. In fact, these points are crucial to many questions related to optimization problems. We will discuss these problems in later pages.

Definition

A function $f(x)$ is said to have a local maximum at c iff there exists an interval I around c such that

$$f(c) \geq f(x) \text{ for all } x \in 1$$

Analogously, $f(x)$ is said to have a local minimum at c iff there exists an interval I around c such that

$$f(c) \leq f(x) \text{ for all } x \in 1$$

A local extremum is a local maximum or a local minimum.

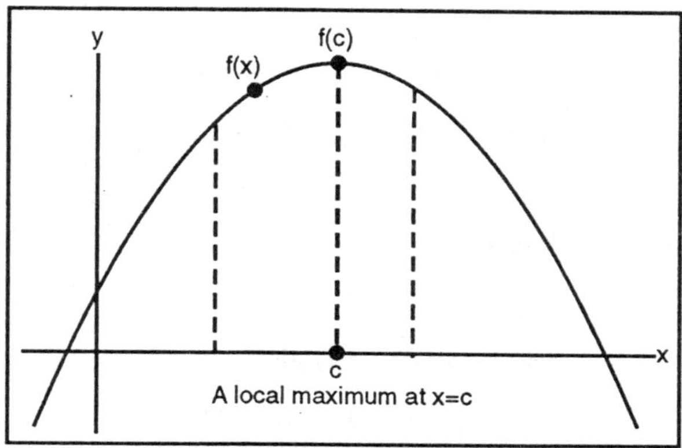

A local maximum at x=c

Using the definition of the derivative, we can easily show that:

If $f(x)$ has a local extremum at c, then either ,

$$f'(c) = 0 \text{ or } f'(c) \text{ does not exist}$$

These points are called critical points.

Example: Consider the function $f(x) = x^3$. Then $f'(0) = 0$ but 0 is not a local extremum. Indeed, if $x < 0$, then $f(x) < f(0)$ and if $x > 0$, then $f(x) > f(0)$.

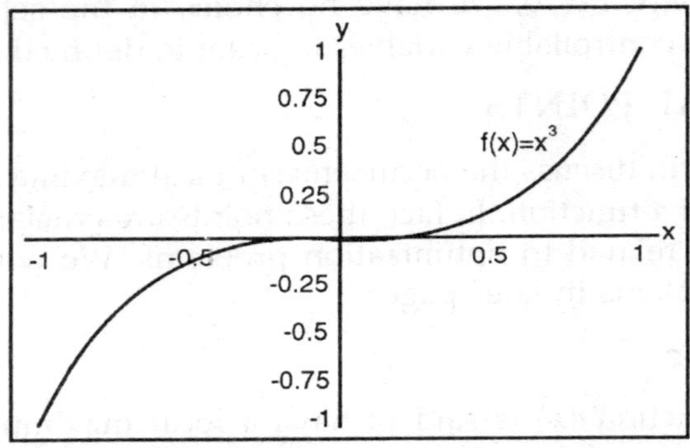

Therefore the conditions

$$f'(c) = 0 \text{ or } f'(x) \text{ does not exist}$$

do not imply in general that c is a local extremum. So a local extremum must occur at a critical point, but the converse may not be true.

Example: Let us find the critical points of,

$$f(x) = |x^2 - x|$$

Answer: We have ,

$$f(x) = \begin{cases} x^2 - x & \text{if } x \le 0 \\ -(x^2 - x) & \text{if } 0 \le x \le 1 \\ x^2 - x & \text{if } 1 \le x \end{cases}$$

Clearly we have,

$$f'(x) = \begin{cases} 2x - 1 & \text{if } x < 0 \\ -2x + 1) & \text{if } 0 < x < 1 \\ 2x - 1 & \text{if } 1 < x \end{cases}$$

Clearly we have,

$$f'(x) = 0 \text{ if } x = \frac{1}{2}$$

Also one may easily show that $f'(0)$ and $f'(1)$ do not exist.

Therefore the critical points are,

Let c be a critical point for $f(x)$. Assume that there exists an interval I around c, that is c is an interior point of I, such that $f(x)$ is increasing to the left of c and decreasing to the right, then c is a local maximum. This implies that if $f'(x) \geq 0$ for $x \geq c$ (x close to c), and $f'(x) \leq 0$ for (x close to c), then$c$ is a local maximum. Note that similarly if $f'(x) \leq 0$ for $x \leq c$ (x close to c), and $f'(x) \geq 0$ for $x \geq c$ (x close to c), then c is a local minimum. So we have the following result:

First Derivative Test. If c is a critical point for $f(x)$, such that $f'(x)$ changes its sign as x crosses from the left to the right of c, then c is a local extremum.

Example: Find the local extrema of,

$$f(x) = |x^2 - x|$$

Answer: Since the local extrema are critical points, then from the above discussion, the local extrema, if they exist, are among the points

$$\frac{1}{2}, 0, 1$$

Recall that,

$$f'(x) = \begin{cases} 2x - 1 & \text{if } x < 0 \\ -2x + 1 & \text{if } 0 < x < 1 \\ 2x - 1 & \text{if } 1 < x \end{cases}$$

(1) For $x = 1/2$, we have

$$\begin{cases} f'(x) > 0 & \text{if } 0 < x < 1/2 \\ f'(x) < 0 & \text{if } 1/2 < x < 1 \end{cases}$$

So the critical point $\frac{1}{2}$ is a local maximum.

(2) For $x = 0$, we have

$$\begin{cases} f'(x) < 0 & \text{if } x < 0 \\ f'(x) > 0 & \text{if } 0 < x < 1/2 \end{cases}$$

So the critical point 0 is a local minimum.

(3) For $x = 1$, we have

$$\begin{cases} f'(x) < 0 & \text{if } 1/2 < x < 1 \\ f'(x) > 0 & \text{if } 1 < x \end{cases}$$

So the critical point -1 is a local minimum.

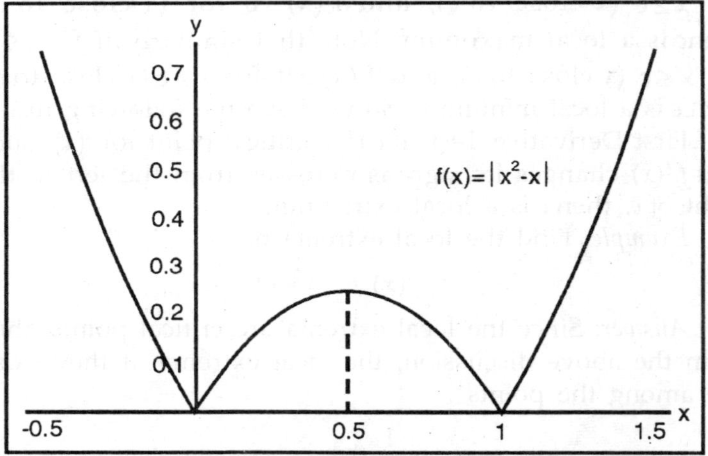

Let c be a critical point for $f(x)$ such that $f'(c) = 0$.

(i) If $f''(c) > 0$, then $f'(x)$ is increasing in an interval around c. Since $f'(c) = 0$, then $f'(x)$ must be negative to the left of c and positive to the right of c. Therefore, c is a local minimum.

(ii) If $f''(c) < 0$, then $f'(x)$ is decreasing in an interval around c. Since $f'(c) = 0$, then $f'(x)$ must be positive to the left of c and negative to the right of c. Therefore, c is a local maximum.

This test is known as the Second-Derivative Test.

Example: Find the local extrema of

$$f(x) = x^5 - 5x.$$

Answer: First let us find the critical points. Since $f(x)$ is a polynomial function, then $f(x)$ is continuous and differentiable everywhere. So the critical points are the roots of the equation $f'(x) = 0$, that is $5x^4 - 5 = 0$, or equivalently $x^4 - 1 = 0$. Since $x^4 - 1 = (x-1)(x+1)(x^2+1)$, then the critical points are 1 and -1. Since $f''(x) = 20x^3$, then

$$f''(1) = 20 > 0 \text{ and } f''(-1) = -20 < 0$$

The second-derivative test implies that $x=1$ is a local minimum and $x=-1$ is a local maximum.

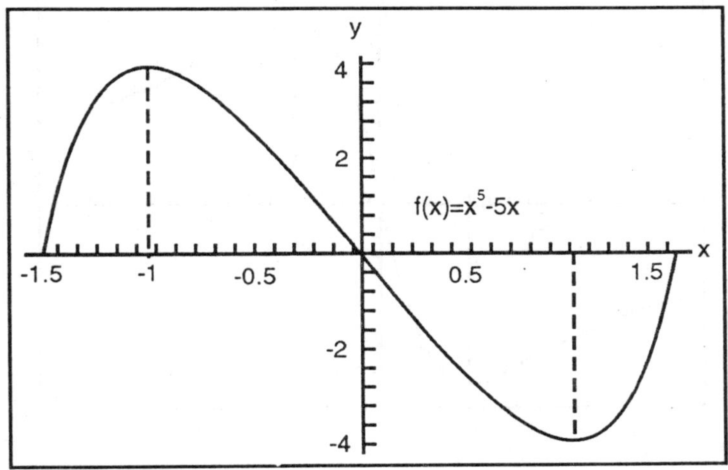

Exercise 1

Find the local extrema of,

$$f(x) = \frac{x}{1+x^2}$$

Answer to Exercise 1

First let us find the critical points. The function $f(x)$ is rational and is defined for any x. In fact, $f(x)$ is differentiable for any x.

Moreover we have,

$$f'(x) = \frac{\left(1+x^2\right) - 2x^2}{\left(1+x^2\right)^2} = \frac{1-x^2}{\left(1+x^2\right)^2}$$

So $f'(x) = 0$ implies $x = \pm 1$. Since

$$\begin{cases} f'(x) < 0 & \text{if } x < -1 \\ f'(x) > 0 & \text{if } -1 < x < 1 \\ f'(x) < 0 & \text{if } 1 < x \end{cases}$$

Then the first-derivative test implies that $x = -1$ is a local minimum and $x = 1$ is a local maximum.

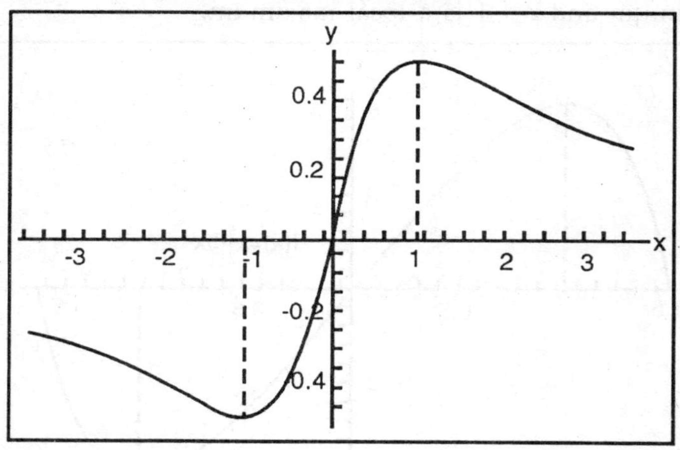

Exercise 2

Find the local extrema of,

$$f(x) = \sin(x) + \cos(x)$$

Answer to Exercise 2

Since sin(x) and cos(x) are continuous and differentiable everywhere, then $f(x)$ is continuous and differentiable everywhere.

So the critical points of $f(x)$ are the roots of,

$$f'(x) = \cos(x) - \sin(x) = 0$$

Hence cos(x)=sin(x) Trigonometric algebra implies that,

$x = \frac{\pi}{4} + 2n\pi$ or $x = \frac{5\pi}{4} + 2n\pi$,

where $n = 0, \pm 1, \pm 2, \ldots$ On the other hand, we have $f''(x) = -\sin(x) - \cos(x)$.

So we have,

$$f''\left(\frac{\pi}{4}+2n\pi\right)=-2\frac{\sqrt{2}}{2}=-\sqrt{2}$$

and

$$f''\left(\frac{5\pi}{4}+2n\pi\right)=2\frac{\sqrt{2}}{2}=\sqrt{2}$$

So the second derivative test implies that,

$$x=\frac{\pi}{4}+2n\pi$$

are local maximum points and,

$$x=\frac{5\pi}{4}+2n\pi$$

are local minimum points.

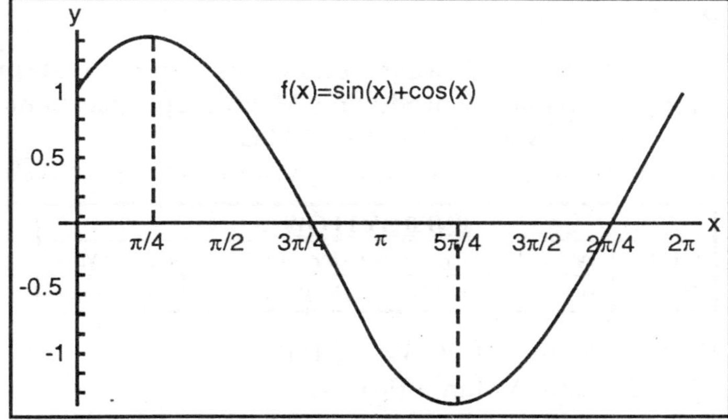

Note that the second derivative test was quite easy to use. In this case it is harder to use the first derivative test. It is sometimes hard to tell in advance which of the two tests to use. Just try one of them, and switch to the other one, if you get stuck!

INTERNAL ENERGY AND ENTROPY OF STEAM

INTERNAL ENERGY

Internal energy is defined as the energy associated with the random, disordered motion of molecules. It is separated in scale from the macroscopic ordered energy associated with moving objects; it refers to the invisible microscopic energy on the atomic and molecular scale. For example, a room temperature glass of water sitting on a table has no apparent energy, either potential or kinetic . But on the microscopic scale it is a seething mass of high speed molecules traveling at hundreds of metres per second. If the water were tossed across the room, this microscopic energy would not necessarily be changed when we superimpose an ordered large scale motion on the water as a whole. U is the most common symbol used for internal energy.

WORK - W, HEAT - Q, AND INTERNAL ENERGY – U

WORK *W*

Useful Energy Transfered across the System's Boundaries, capable of producing Macroscopic-Mechanical Motion of a the system's Centre-of-Mass.

W = Work done by (or on) one system on another system

ENERGY FLOW			
OUT:	$W > 0$	*System Does External Work*	*Sys --> Work*
INTO:	$W < 0$	*Work Done on the System*	*Work --> Sys*

Work done by a Gas : $W = \int_0^f PdV$

W=0 Constant Volume Process

W=PΔV Constant Pressure Process, $W = PV\ln\frac{V_f}{V_0}$

Constant Temperature Process $W = \frac{P_0V_0 - P_fV_f}{\gamma - 1}$

Adiabatic Process Q = 0

HEAT Q

Energy Transfer across the System's Boundaries that cannot produce Macroscopic-Mechanical Motion of the system's Centre-of-Mass. Energy Transfer at the Molecular Level .

Q = Microscopic Energy flow into (or out of) the System,

ENERGY FLOW			
INTO:	$Q > 0$	*System Absorbs Heat*	*Heat --> Sys*
OUT:	$Q < 0$	*System Releases Heat*	*Sys --> Heat*

Some common types of Heat,

Lost: $Q = mC\Delta T \pm mL_f \pm mL_v$

Solids or Liquids $Q = mC_p\Delta T$

Gas- Constant Pressure Process $Q = mC_v\Delta T$

Gas - Constant Volume Process,

$$Q = PV \text{ In} \frac{V_f}{V_0}$$

Gas - Constant Temperature Process $Q = 0$

Gas - Adiabatic Process

Internal Energy U

Energy Stored in a System at the Molecular Level. The System's Thermal Energy -the Kinetic Energy of the atoms due to their random motion relative to the Centre of Mass plus the binding energy (Potential Energy) that holds the atoms together.

U = Microscopic Energy contained in the System

MICROSCOPIC ENERGY

Internal energy involves energy on the microscopic scale. For an ideal monoatomic gas, this is just the translational kinetic energy of the linear motion of the "hard sphere" type atoms, and the behaviour of the system is well described by kinetic theory.

However, for polyatomic gases there is rotational and vibrational kinetic energy as well. Then in liquids and solids there is potential energy associated with the intermolecular attractive forces.

A simplified visualization of the contributions to internal energy can be helpful in understanding phase transitions and other phenomena which involve internal energy.

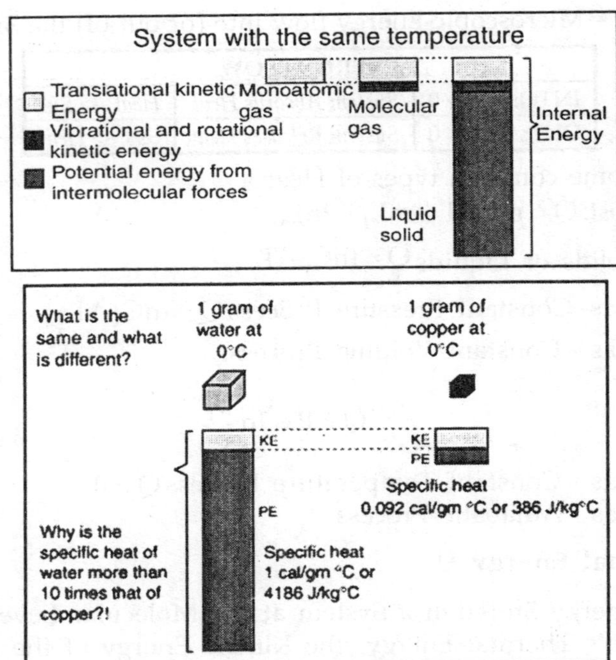

Internal Energy

When the sample of water and copper are both heated by 1°C, the addition to the kinetic energy is the same, since that is what temperature measures. But to achieve this increase for water, a much larger proportional energy must be added to the potential energy portion of the internal energy. So the total energy required to increase the temperature of the water is much larger, i.e., its specific heat is much larger.

A heavy ball with an initial kinetic energy of 4000 J is trapped inside a box with rigid walls containing a cylinder constructed of small light-weight spheres. The ball crashes into the cylinder and breaks it apart. The bar graph at the right and the table at the bottom display the kinetic energy of the ball.

ENTROPY OF STEAM

The entropy diagram for steam is often convenient because it shows the relationship between

- Pressure
- Temperature
- Dryness Fraction
- Entropy

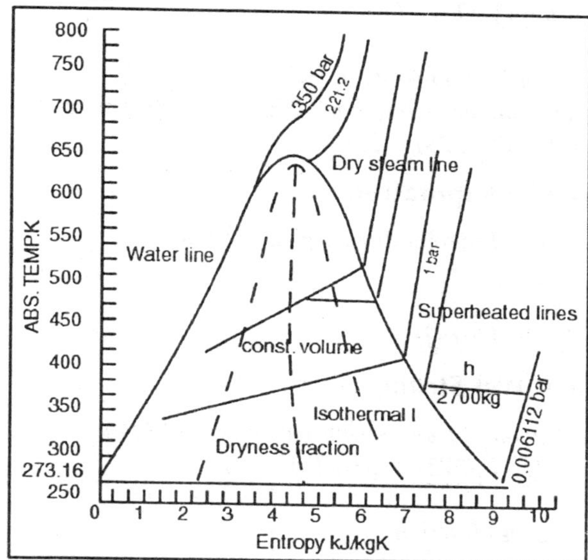

With two of the factors given - the others can be found in the diagram. The ordinates in the diagram represents the Entropy and the Absolute temperature.

The diagram consist of the following lines

- Isothermal line
- Pressure lines
- Lines of dryness fraction
- Waterline between water and steam
- Dry steam lines
- Constant volume lines

The total heat is given by the area enclosed by absolute zero base water line and horizontal and vertical line from the respective points. An adiabatic expansion is a vertical line. An adiabatic process is expansion at constant entropy with no transfer of heat.

- Critical temperature of steam is *375 to 3,380°C*
- Critical pressure is *217.8 atm*

Total Entropy of Steam

Entropy of Water

The change of entropy can be expressed as:
$dS = log_e(T_1/T)$ *(1)*
where,
T *= absolute temperature (K)*
The entropy of water above freezing point can be expressed as:
$dS = log_e(T_1/273)$ *(2)*

Entropy of Evaporation

Change of Entropy during evaporation:
$dS = dL/T$ *(3)*
where
L *= latent heat (J)*

Entropy of Wet Steam

The entropy of wet steam can be expresses as:
$dS = log_e(T_1/273) + \zeta(L_1/T_1)$ *(4)*
where
ζ *= dryness fraction*

Entropy of Superheated Steam

Change of entropy during superheating can be expressed as:
$dS = c_p\, log_e(T/T_1)$ *(5)*
where
c_p *= specific heat capacity at constant pressure for steam (kJ/kgK)*
The entropy of superheated steam can be expressed as:
$dS = log_e(T_1/273) + L_1/T_1 + c_p\, log_e(T_s/T_1)$ *(6)*
where
T_s *= absolute temperature of superheated steam*
T_1 *= absolute temperature of evaporation*

Entropy of Superheated Steam (kJ/kgK)

The saturated steam is exposed to a surface with a higher temperature, its temperature will increase above the evaporating temperature. The steam is then described as superheated by the temperature degrees above saturation temperature.

Absolute pressure		Saturation temperature	Steam Temperature (°C)						
(kN/ m²)	(bar)	(°C)	120	150	180	200	230	250	280
150	1.5	111.4	7.239	7.419	7.557	7.644	7.767	7.845	7.957
200	2	120.2		7.279	7.420	7.507	7.631	7.710	7.822
250	2.5	127.4		7.169	7.311	7.400	7.525	7.604	7.717
350	3.5	138.9		6.998	7.146	7.237	7.364	7.444	7.558
400	4	143.6		6.929	7.079	7.171	7.299	7.380	7.495
500	5	151.8			6.965	7.059	7.190	7.272	7.388
600	6	158.8			6.869	6.966	7.100	7.183	7.300
700	7	165.0			6.786	6.886	7.022	7.107	7.225
800	8	170.4			6.712	6.815	6.954	7.040	7.156
900	9	175.4			6.645	6.751	6.893	6.980	7.101
1000	10	179.9			6.584	6.692	6.838	6.926	7.049
1100	11	184.4				6.638	6.787	6.876	7.001
1200	12	188.0				6.587	6.739	6.831	6.956
1400	14	195.0				6.494	6.653	6.748	6.877
1600	16	201.4					6.577	6.674	6.806
2000	20	212.4					6.440	6.546	6.685
2500	25	223.9					6.292	6.407	6.558
3500	35	242.5						6.173	6.349

- Note! Steam cannot be superheated whilst it is still in the contact with water, because additional heat will evaporate more water, cooling down the superheated steam.

Superheated steam is produced by passing saturated steam through an additional heat exchanger.

Superheated steam is also called:

- Surcharged steam
- Anhydrous steam
- Steam gas

USE OF STEAM TABLES

PROPERTY DIAGRAMS AND STEAM TABLES

If the substance is not water vapour, the "state" of the substance is usually obtained through the use of T-s (temperature-entropy) and h-s (enthalpy-entropy) diagrams, available in most thermodynamics texts for common substances. The use of such diagrams is demonstrated by thefollowing two examples.

Example: Use of the h-s diagramMercury is used in a nuclear facility. What is the enthalpy of the mercury if its pressureis 100 psia and its quality is 70%?

Solution: From the mercury diagram, Figure A-3 of Appendix A, locate the pressure of 100 psia.Follow that line until reaching a quality of 70%. The intersection of the two lines givesan enthalpy that is equal to h = 115 Btu/lbm.

HEATING BOILER EXPANSION TANKS

Heating boiler expansion tanks are installed to absorb the initial pressure that occurs when the heating system warms up. Air molecules entrained in water inside the heatin boiler itself as well as in the heating system piping, baseboards, or radiators, expands and thus cause an initial pressure increase in the heating system.

If we do not provide some way to absorb this initial pressure increase, it is possible that the heating system's internal pressure would exceed 30 psi - the typical point at which a heating boiler pressure/temperature relief valve will open to spill excess pressure. If the relief valve is forced open in this manner the heating system will first lose water each time a heating cycle starts by heating up the boiler.

Then the heating system will take in makeup water (through the automatic water feed valve) each time the system cools down. The result would be recurrent loss and then inflow of water through the boiler, increasing the risk of system corrosion as well as wasting water and possibly causing other damage or operating problems.

When the heating boiler and system are cool, the expansion tank will contain mostly air. As the heating system warms up and air entrained in the water raises system pressure, the increased pressure forces some of the heating system water into the expansion tank, thus permitting the tank to absorb the *initial* increase in system pressure.

Naturally if the heating system is not operting properly and the boiler temperature or pressure continue to rise above acceptable levels, and since the expansion tank is a fixed volume which cannot absorb an unlimited heating system pressure increase, eventually the relief valve should also open to spill excess pressure.

TROUBLE WITH HEATING BOILER EXPANSION TANK

This means that if you see dripping at the pressure/temperature relief valve on a heating boiler, one thing to check is whether or not the expansion tank is working properly. We discuss relief valve leaks caused by expansion tank problems just below.

Expansion tanks on hot water heating systems can be divided roughly into two groups: older type bladderless heating system expansion tanks and newer type bladder-type expansion tanks such as those sold by Extrol(R).

HEATING SYSTEM EXPANSION TANKS

Older heating system expansion tanks such as the one shown in this photo need periodic service: because air in the expansion tank can become absorbed into the heating water over time, eventually the expansion tank can become waterlogged.

A waterlogged heating system expansion tank will be unable to absorb the initial heating system pressure increase - it stops working. This can lead to spillage at the boiler relief valve each time the heating system warms up.

Spillage at the relief valve is potentially dangerous: eventually minerals in the water can clog a leaky relief valve, causing it to stop leaking - which might look ok but this means that the relief valve has become clogged - the boiler is operating without this critical safety device and an explosion could occur.

We can determine when the expansion tank needs drainage: If the tank is heavy (try pushing it up or tapping on it) or if the relief valve is leaking, we probably need to drain the tank and let air return to it. Many expansion tanks use a special drain valve that permits air to flow into the tank as water is drained out. We drain the expansion tank simply using a garden hose connected to its drain valve. But watch out, the draining water could be hot.

Newer expansion tanks using an internal bladder should not need service. Newer type heating system expansion tanks that use an internal bladder keep their water and air separated. These tanks should not need service. If a bladder-type expansion tank has become waterlogged it's because the bladder has ruptured and the tank needs repair or replacement.

HEATING UP WITH STEAM

The amount of heat required to raise the temperature of a substance can be expressed as:

$Q = m\, c_p\, dT \quad (1)$

where

Q = quantity of energy or heat (kJ)

m = mass of the substance (kg)

c_p *= specific heat capacity of the substance (kJ/kg °C) - Material Properties and Heat Capacities for several materials*

dT = temperature rise of the substance (°C)

Preferring Imperial Units - Use the Units Converter!

This equation can be used to determine a total amount

of heat energy for the whole process, but it does not take into account the rate of heat transfer which is amount of heat energy per unit time

In non-flow type applications a fixed mass or a single batch of product is heated. In flow type applications the product or fluid is heated when it constantly flows over a heat transfer surface.

NON-FLOW OR BATCH HEATING

In non-flow type applications the process fluid is kept as a single batch within a tank or vessel. A steam coil or a steam jacket heats the fluid from a low to a high temperature.

The mean rate of heat transfer for such applications can be expressed as:

$q = m\ c_p\ dT/t$ (2)

where

q = *mean heat transfer rate (kW (kJ/s))*

m = *mass of the product (kg)*

c_p = *specific heat capacity of the product (kJ/kg.°C) - Material Properties and Heat Capacities for several materials*

dT = *Change in temperature of the fluid (°C)*

t = *total time over which the heating process occurs (seconds)*

FLOW OR CONTINUOUS HEATING PROCESSES

In heat exchangers the product or fluid flow is continuously heated.

The mean heat transfer can be expressed as:

$q = c_p\ dT\ m/t$ (3)

where

q = *mean heat transfer rate (kW (kJ/s))*

m/t = *mass flow rate of the product (kg/s)*

c_p = *specific heat capacity of the product (kJ/kg.°C) - Material Properties and Heat Capacities for several materials*

dT = *change in temperature of the fluid (°C)*

CALCULATING THE AMOUNT OF STEAM

If we know the heat transfer rate - the amount of steam can be calculated:

$m_s = q/h_e$ (4)

where

m_s = *mass of steam (kg/s)*

q = *calculated heat transfer (kW)*

h_e = *evaporation energy of the steam (kJ/kg)*

EXAMPLE - BATCH HEATING BY STEAM

A quantity of water is heated with steam of 5 bar from a temperature of *35 °C to 100 °C* over a period of *20 minutes (1200 seconds)*. The mass of water is *50 kg* and the specific heat capacity of water is *4.19 kJ/kg.°C*.

Heat transfer rate:

$q = (50\ kg)\ (4.19\ kJ/kg\ °C)\ (100°C - 35°C)/(1200\ s)$

$= 11.35\ kW$

Amount of steam:

$m_s = (11.35\ kW)/(2085\ kJ/kg)$

$= 0.0055\ kg/s$

$= 19.6\ kg/h$

EXAMPLE - CONTINUOUSLY HEATING BY STEAM

Water flowing at a constant rate of *3 l/s* is heated from *10 °C to 60 °C* with steam at *8 bar*.

The heat flow rate can be expressed as:

$q = (4.19\ kJ/kg.°C)\ (60°C - 10°C)\ (3\ l/s)\ (1\ kg/l)$

$= 628.5\ kW$

The steam flow rate can be expressed as:

$m_s = (628.5\ kW)/(2030\ kJ/kg)$

$= 0.31\ kg/s$

$= 1115\ kg/h$

MEASUREMENT OF DRYNESS FRACTION

To produce *100%* dry steam in an boiler, and keep the steam dry throughout the piping system, is in general not possible. Droplets of water will escape from the boiler surface. Because of turbulence and splashing when bubbles of steam break through the water surface the steam space will contain a mixture of water droplets and steam.

In addition heat loss in the pipes will condensate steam to droplets of water. Steam produced in a boiler where the heat is supplied to the water and where the steam are in contact with the water surface of the boiler - will contain approximately 5% water by mass.

DRYNESS FRACTION OF WET STEAM

If the water content of the steam is 5% by mass, then the steam is said to be 95% dry and has a dryness fraction of 0.95.

Dryness fraction can be expressed as:

$\zeta = w_s / (w_w + w_s)$ (1)

where

ζ = *dryness fraction*

w_w = *mass of water (kg, lb)*

w_s = *mass of steam (kg, lb)*

ENTHALPY OF WET STEAM

The actual enthalpy of evaporation of wet steam is the product of the dryness fraction - ζ - and the specific enthalpy- h_s - from the steam tables. Wet steam have lower usable heat energy than dry saturated steam.

$h_t = h_s \zeta + (1 - \zeta) h_w$ (2)

where

h_t = *enthalpy of wet steam (kJ/kg, Btu/lb)*

h_s = *enthalpy of steam (kJ/kg, Btu/lb)*

h_w = *enthalpy of saturated water or condensate (kJ/kg, Btu/lb)*

SPECIFIC VOLUME OF WET STEAM

The droplets of water in wet steam will occupy negligible space in the steam and the specific volume of wet steam will be reduced according the dryness fraction.

$v = v_s \zeta$ (3)

where

v = *specific volume of wet steam (m^3/kg, ft^3/lb)*

v_s = *specific volume of the dry steam (m^3/kg, ft^3/lb)*

ENTHALPY AND SPECIFIC VOLUME OF WET STEAM

Steam at pressure *5 bar gauge* has a dryness fraction of *0.95*.

Total enthalpy can be expressed as:

h_t = (2085 kJ/kg) 0.95 + (1 - 0.95) (670.4 kJ/kg)

= 2,014 kJ/kg

Specific volume can be expressed as:

$v = (0.315\ m^3/kg)\ 0.95$

$= 0.299\ m^3/kg$

DRYNESS FRACTION VARIATION

The density of a cubic metre of wet steam is higher than that of a cubic metre of dry steam. If the quality of steam is not taken into account as the steam passes through the flowmetre, then the indicated flowrate will be lower than the actual value.

To reiterate; dryness fraction is an expression of the proportions of saturated steam and saturated water. For example, a kilogram of steam with a dryness fraction of 0.95, contains 0.95 kilogram of steam and 0.05 kilogram of water.

Example: As a basis for the following examples, determine the density of dry saturated steam at 10 bar g with dryness fractions of 1.0 and 0.95.

Dryness Fraction(x) = 1.0

Specific Volume of dry steam (V_g)

at 10 bar g (from steam tables) = $0.1773\ m^3/Kg$

$$\text{Density } (\rho) = \frac{1}{0.1773\,\text{m}^3/\text{Kg}}$$

with x having a dryness fraction

of 1.0, density (ρ) = $5.6414 Kg/m^3$

Dryness fraction (x) = 0.95

Specific volume of dry steam (V_g)

at 10 bar g (from steam tables) = $0.1773\ m^3/Kg$

Spcific Volume of Water (V_f)

at 10 bar g (from steam tables) = 0.0011

- Volume occupied by steam @x = 0.95 = 0.95×0.1773 = 0.1684 m^3
- Volume occupied by Water @x = 0.95 = 0.05×0.0011 = 0.000055m^3
- Total volume occupied by steam and water = 0.1684 + 0.000055m^3 = 0.168455m^3

Density (ρ) of mixture,

$$= \frac{1}{0.168455\,m^3/Kg} = 5.9363\,Kg/m^3$$

Difference in density = 5.9363 Kg/m^3 – 5.6414 Kg/m^3 = 0.2949 Kg/m^3

Therefore, a reduction in volume is calculated to be 4.97%,

$$\text{Density of steam} = \frac{1}{v_g x}$$

$$\frac{\text{Indicated massflowrate}}{\text{Actual flowrate}} = \frac{\sqrt{\text{Density at calibrated dryness fraction}}}{\sqrt{\text{Density at actual dryness fraction}}}$$

Equation

The proportion of the volume occupied by the water is approximately 0.03% of that occupied by the steam. For most practical purposes the volume occupied by the water can be ignored and the density of wet steam can be defined as shown in Equation

Where:

Vg = Specific volume of dry steam

x = Dryness fraction

Using Equation 4.4.3, find the density of wet steam at 10 bar g with a dryness fraction (x) of 0.95. The specific volume of dry steam at 10 bar g (g) = 0.1773 m3/kg. This compares to 5.9363 kg/m3 when calculated as a mixture.

Effect of Dryness Fraction on Flowmetres

To reiterate earlier comments regarding differential pressure flowmetre errors, mass flowrate (qm) will be proportional to the square root of the density, and density is related to the dryness fraction. Changes in dryness fraction will have an effect on the flow indicated by the flowmetre.

Equation can be used to determine the relationship between actual flow and indicated flow:

$$\frac{\text{Indicated flowrate}}{\text{Actual flowrate}} = \frac{\sqrt{\text{Density at } x = 1.00}}{\sqrt{\text{Density at } x = 0.95}}$$

$$\frac{1 \text{ Kg/s}}{\text{Actual flowrate}} = \frac{\sqrt{5.6414}}{5.9294} = \frac{2.3752}{2.4350}$$

Actual flowrate = 1.0252 Kg/s

$$\text{Per centage error} = \frac{\text{Indicated flow} - \text{Actual flow}}{\text{Actual flow}} \times 100\%$$

$$\text{Per centage error} = \frac{1 - 1.0252}{1.0252} \times 100\% = -2.46\%$$

Equation: All steam flowmetres will be calibrated to read at a pre-determined dryness fraction, the typically value is 1. Some steam flowmetres can be recalibrated to suit actual conditions.

Example: Using the data from Example determine the per centage error if the actual dryness fraction is 0.95 rather than the calibrated value of 1.0, and the steam flowmetre was indicating a flowrate of 1 kg/s. Therefore, the negative sign indicates that the flowmetre under-reads by 2.46%.

Equation is used to compile the graph shown in Figure:

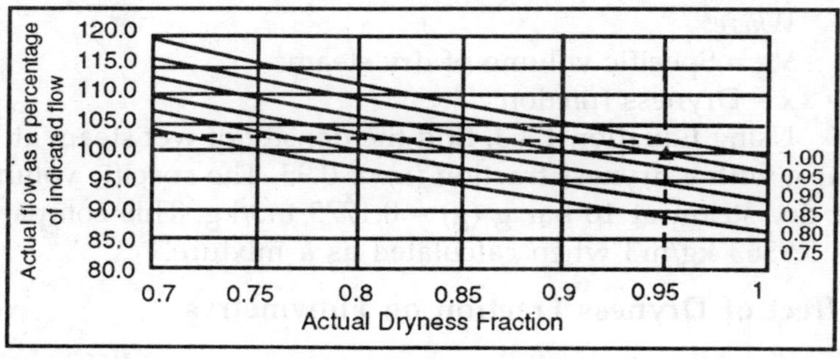

Fig. Effect of Dryness Fraction on Differential Pressure Flowmetres

Effect of Dryness Fraction on Vortex Flowmetres

It can be argued that dryness fraction, within sensible limitations, is of no importance because:

- Vortex flowmetres measure velocity.
- The volume of water in steam with a dryness fraction of, for example, 0.95, in proportion to the steam is very small.
- It is the condensation of dry steam that needs to be measured.

However, independent research has shown that the water droplets impacting the bluff body will cause errors and as vortex flowmetres tend to be used at higher velocities, erosion by the water droplets is also to be expected. Unfortunately, it is not possible to quantify these errors.

Chapter 5

Boilers

STEAM BOILERS

Boiler classification can be based on many factors like usage, fuel fired, fuel firing system, type of arrangement etc. Commonly known types are pulverized coal fired boilers, fluidized bed boilers, super critical boilers, oil and gas fired boilers. All cater to industrial and power generation.

DEFINITION

A boiler can be defined as a closed vessel in which water or other fluid is heated under pressure. This fluid is then circulated out of the boiler for use in various processes or power generation. In the case of power generation steam is taken out of the steam boiler at very high pressure and temperature.

In 200 B.C. a Greek named Hero designed a very simple machine which used the steam, generated in a vessel heated from below, to rotate a wheel as the steam escaped through two small pipes kept diametrically opposite, he called it as *Aelopile*.

CLASSIFICATION

Boilers vary considerably in detail and design.Most boilers may be classified and described interms of a few basic features or characteristics.Some knowledge of the methods of classificationprovides a useful basis for understanding thedesign and construction of the various types ofnaval boilers.In the following paragraphs, we haveconsidered the classification of naval boilersaccording to intended service, location of fire andwater spaces, type of circulation, arrangement ofsteam and water spaces, number of furnaces,burner location, furnace pressure, type of super-heaters, control of superheat, and operatingpressure

From 200 B.C. to date, many developments have taken place that today allow us to classify steam boilers in different ways.

Hence steam generating boilers can be classified under various categories. The main purpose of steam boilers is to generate steam, and so the way in which the steam is generated and consumed forms the major category.

The major two groups of boiler application are Industrial steam generators and power generation boilers. Boilers are also classified as fire tube and water tube boilers.

Fire tube boilers have almost become extinct; however this can be classified as:

- Locomotive boilers, which ruled rail transportation before diesel and electric engine came.
- Industrial boilers, mainly used for green projects where initial steam is required
- Domestic use boilers

Water tube boilers took over when size and capacity increased. This can be classified depending on type of circulation used to generate steam as:

- Natural circulation boiler
- Forced circulation boilers
- Super critical pressure boilers or zero circulation boilers

Depending on type of firing adopted in boilers they can be classified as:

- Stoker fired
- Pulverized coal fired
- Down shot fired
- Fluidized bed boilers
- Cyclone fired
- Chemical recovery boilers
- Incinerators

Of these the stokers which were predominantly used in early days of high pressure high capacity boilers are being replaced by pulverized coal fired boilers and fluidized bed boilers.

Stoker boilers are still designed and used in few applications like sugar industries, etc. Fluidized boilers are also going through fast development and can be now sub classified as

- Bubbling fluidized bed boilers
- Pressurized fluidized bed boilers
- Circulating fluidized bed boilers.

The higher capacity boilers are mainly circulating fluidized bed boilers due inherent limitations in bubbling bed boilers.

Boilers can be classified based on the type of fuel used as:

- Coal fired boilers
- Oil fired boilers
- Gas fired boilers
- Multi-fuel fired
- Industrial waste fired boilers
- Biomass fired boilers

Various types of arrangement are used by designers in designing the boiler for meeting the end requirement. Hence boilers are classified based on the arrangement as

- Top supported boilers
- Bottom supported
- Package boilers
- Field erected boilers
- Drum type boilers
  - Single drum
  - Bi drum
  - Three drums, but these are presently out of use
- Tower type or single pass
- Close coupled
- Two pass boilers

Boilers therefore can be classified based on firing type, fuel used, construction type, circulation type, firing system design nature, and nature of steam application.

Today's steam generating systems owe their dependability and safety to more than 125 years of experience in the design, fabrication, and operation of water tube boilers.

LANCASHIRE AND LOCOMOTIVE BOILERS

LANCASHIRE BOILERS

All the boilers at Pleasley were fire-tube shell types. At Teversal colliery a few years earlier, externally fired egg-ended boilers had been installed but at Pleasley it was decided to install Cornish and Lancashire boilers.

It's not certain why two different types were installed especially as the former were less efficient. It could be that the Cornish boilers were adequate for the demands of the screens and the South-pit winder which, in the early days, was not used for coal-winding, whilst the coal winding demands of the North winder would have been met by the Lancashire boilers. The Cornish ones would later be replaced by Lancashire boilers.

By 1892 there were four sets of main boilers, running at 50 psi., located on either side of the chimneys on each side of both engine-houses and feeding the winders, the screen engines and the workshop engines.

In addition, there were another four running at 50 psi, located near to the ventilation fans, feeding the fan and the dynamo engines.

The Cornish boilers were 40 ft long but the Lancashire ones were only 30 ft. A sketch plan made by the colliery manager showing the North engine-house and boilers in 1901 indicates that the boilers were arranged perpendicular to the engine-houses and that the Cornish boilers were located on the west side.

In 1900, the coal winding facilities at the South pit were upgraded and a more powerful winder was installed. The 4 Lancashire boilers in the south-east range were replaced by 5 new ones. This time they were aligned parallel with the engine-house and were flued into the side of the chimney. The coal supply was tipped from sided-door wagons on a short tightly curved railway siding.

These new boilers were the recently introduced Thompson dish-ended types. They were insulated by 5 layers of bricks, but the boiler-house itself was not enclosed when the photo was taken - some time between 1900 and 1905. They were raised above the boiler-house floor with the fire-doors at about head height and firing must have been quite tricky. It is not known whether these boilers were fitted with Galloway flue tubes.

The engines supplied by these boilers were said to operate at 100 psi and they were protected by dead-weight safety-valves carrying ten plates. The pressure-gauge on the nearest boiler, however, which is venting steam from the safety-valve, although somewhat indistinct, seems to be reading at about the 10 oclock position. This seems rather low - unless the guages read up to 300 psi - which seems rather high since the boilers are unlikely to have been rated at more than 150 psi. They do not seem to have been fitted with low-water safety valves which is rather surprising. The water-level sight glasses were not protected by glass casing either - this only became mandatory in 1911.

The feed water was supplied along insulated pipes and the steam main was also well insulated although the flanges were unprotected. There was no superheating of the steam. The boilers were hand fired and the ash disposal arrangements are not obvious A new boiler-house was constructed on the east side of the South engine-house and 9 new Lancashire boilers were installed. Three existing boilers in the North engine-house east-side boiler house were retained, giving a consolidated range of twelve. Twin Unit superheaters were fitted at the rear of each flue tube giving a superheat of 90 deg F. The boilers were hand fired but it's not clear what the coal delivery and ash removal arrangements were.

Some of the dish-ended boilers installed in 1900 were transferred to Teversal and Silverhill collieries and some to the Stanton ironworks itself. Other, older boilers, were converted into exhaust-steam accumulators for the turbines, the remainder being removed altogether. A few years later, the fan-engine boilers were also removed and transferred to Silverhill colliery.

This boiler arrangement lasted until the colliery was closed. The ash removal underwent several modifications in later years and the firing arrangements were upgraded to mechanical stoking from overhead bunkers in the 1950s upgrade.

LOCOMOTIVE BOILER

It is intended that this short course will assist train crews to improve their operational skills on the foot plate by making them more aware of the capabilities of the boiler and to give a better understanding of the various parts and controls. The course will cover general boiler construction, boiler types and the various mountings attached to them with a description of their construction and correct operation. An explanation of the physical properties of steam and the physical and chemical properties of the principle fuels, coal and fuel oil. An explanation of the principles of combustion and the transformation of heat into power.

OPERATION OF LOCOMOTIVE BOILERS AND FITTINGS

The requirements of a locomotive boiler are very exacting, it having to withstand high steam pressures with a large margin of safety, combined with efficiency and economy of space. The restricted space at the disposal of the designers, and the large area of heating surface required have determined the form of the boiler. The efficiency of a boiler is measured by the amount of water that can be evaporated per lb. of coal, and this depends on the quantity of coal that can be consumed on the firegrate. For the economical production of steam, therefore, a boiler must be well designed for its work and must be handled with a degree of intelligence.

TYPES OF BOILERS AND FIREBOXES

The boiler or steam generator consists essentially of the steel shell, which includes the boiler barrel, the outer firebox wrapper plate, back plate, throat plate and smokebox tubeplate. To this is fitted the inner firebox and steel flue and smoke tubes.

Boiler for supplying superheated steam. The latter diagram illustrates a taper boiler, the cylindrical barrel is made in two sections with the larger diametre at the rear, were the barrel is joined to the outer firebox. The dome in this design, which houses the regulator valve and the auxiliary internal steam pipes, is positioned on top of the rear sloping section of the barrel, where it forms a collector for the steam above the surface of the water.

Fireboxes may be of the deep, long narrow type between the frames or of the shallow, wide type, for example, as fitted to 4-6-2 classes of locomotives. In the latter case the firebox is spread over the frames. The wide type of firebox is fitted when a large grate area is necessary.

The inner firebox is supported from the outer firebox by the foundation ring at the bottom, by crown stays at the top, and by palm stays between the firebox tubeplate and the boiler barrel. In addition, the firebox and outer wrapper plates, backplate and throatplate are stayed together with steel or copper stays, at about 4 in pitch. There are over a thousand of these stays in every locomotive boiler. Longitudinal stays are fitted between the boiler backplate and the smokebox tubeplate, and cross stays between the firebox sides above the crown. From the firebox tubeplate, the steel flue tubes, which may be anything from 1.1/2 to 2.1/4 in. diametre, pass through the boiler barrel to the smokebox tubeplate. When the boiler is fitted with a superheater, a number of large flue tubes (approx. 5 in. dia) are provided in which the superheater elements are positioned.

Some boilers employ the flat top or 'Belpaire' firebox, this being designed to give a larger heating surface and a greater steam space over the firebox. The other type in general use is the round top firebox fitted to earlier designs although these were preferred by the Eastern region of British Railways until the. end of steam. It is normal practice in this country for the inner fireboxes to be made of COPPER The locomotives of the Southern region, are, however fitted with steel fireboxes and thermic syphons which improve the circulation of water around the boiler.

Copper for the inner firebox is used on account of it being a very good conductor of heat, but more especially because of its ductility and suitability for withstanding the great fluctuations of temperature within the firebox.

THE SMOKEBOX

The smokebox is an extension at the front end of the boiler barrel, which together with the blast pipe and the chimney, forms the means of producing an induced draught to the air required for combustion of the fuel in the firebox. Apart from the chimney orifice it is airtight. Other fittings in the smokebox are; The superheater header (when fitted), main steam pipes to the cylinders, Blower and ejector exhaust pipes. On some locomotives, notably Western region types the regulator valve is part of the superheater header.

The Spark Arrestor

This is fitted to the smokebox to prevent the emission of live ashes from the fire being ejected from the chimney when the engine is working hard. On some locomotives it is simply a mesh wire basket fitted between the blast nozzle and the base of the chimney known as the petticoat pipe, the larger particles being arrested and remaining in the smokebox to be removed by hand at the end of the day.

A more complex type of spark arrestor has been fitted to some locomotives that are used on Railtrack lines due to the more stringent spark emission regulations. This has a vertical diaphragm plate at the back of the smokebox that breaks up the larger pieces and then deflects them towards the front of the smokebox were the flow of air is much less, the ashes are then drawn through the wire net screen before being ejected through the chimney, by which time they are small, dead and harmless.

The Superheater

The superheater is the part of the boiler designed to produce superheated steam, that is steam that is heated further after leaving the boiler and out of contact with the water from which it was generated. It is then fed directly to the cylinder valve chests. It consists of a steam collector or header for distributing steam from the boiler via the regulator valve to a series of small tubes called ELEMENTS which pass through the larger flue tubes and return to the header as superheated steam. The steam on passing through these elements has absorbed heat from the hot gases from the firebox on their way through the flue tubes to the smokebox and out of the chimney.

The superheater header is attached to the smokebox tubeplate at the outlet of the main internal steam pipe from the regulator valve and is placed horizontally across the top of the smokebox. At each side of the header are flanges for connection of the main steam pipes to the cylinders. see diagram.

The Elements are fabricated from lengths of solid drawn steel tubing with three return bends, clipped together to form a bundle or element and are attached to the header with a ball and socket joint. These are known as "Melesco" elements after the company which designed them. The number of elements will vary with the degree of superheat required, however all the steam that passes through the regulator valve has to pass through the superheater elements before it can reach the cylinders.

The Brick Arch

The brick arch as its name suggests is made from high temperature refractory material, and is constructed within the firebox. It extends from the tubeplate just below the bottom row of tubes and is inclined upward for just over half the length of the firebox.

Its purpose is to lengthen the path the hot gasses have to take allowing more complete combustion of the volatile matter given off from the fuel on the firebed. When the refractory reaches its normal temperature (white hot over 2500°F) from the radiation of the firebed this further assists in the combustion of the volatile matter given off by the fuel. Secondary air passing through the firehole door is directed under the brick arch by the deflector plate and allowed to mix with the gasses on the firebed and so improving combustion.

Lastly the brick arch allows the gasses to spread evenly over the tubeplate allowing the boiler to absorb as much heat as possible before being passed up the chimney.

The Firehole Door And The Deflector Plate

Various patterns of firehole door are fitted to locomotives, these give access for firing and are a method of controlling the secondary air to the firebox. Large amounts of secondary air entering the firebox is harmful to the boiler and can result in leaking tubes and stays.

Deflector Plate Or Baffle Plate

This is a metal scoop formed to the profile of the firehole and extends for about 20ins into the firebox. It is inclined downwards on the same plane as the brick arch and should point under it. A deflector plate that points over the brick arch is totally useless and must not be used. Its principle purpose is to deflect the cold incoming air under the brick arch which is white hot and preheat the air whilst allowing the air to mix with the gasses on the firebed and so assisting proper combustion.

As stated above it goes a long way to preventing cold air impinging on the boiler plates causing steep temperature changes and causing leakage. Finally in the serious event of low water level and a fusible plug melting the deflector plate will prevent serious eruption of steam back through the firehole door. A correct fitting deflector plate must always be in place when the boiler is in steam.

Drop Grates, Rocking Grates And The Hopper Ash Pan

Most of our locomotives are now fitted with drop grates and hopper ashpans and are far better than the older plain firebar grates, they are there to facilitate disposal of the fire. They are of various types but the most common being the BR STD. type These consist of hinged firebars which can be controlled from the cab by means of levers. A two way stop and locking plate enables the grates to be operated with a limited amount of movement so as to break up the clinker when running, or to be rocked fully to enable the fire to be dropped completely at disposal.

A hopper ashpan is provided on most locomotives and this has hopper doors at the bottom. The doors are held shut with a catch on the side of the ashpan and are operated with the same lever as the rocking grate. The loco must not be run with the doors open and care should be taken to see that the catch is secure before moving off shed. 1 The hopper doors should always be opened before dropping the fire during disposal so as to prevent the hot fire doing serious damage to the ashpan. This operation should be done over an ashpit or other authorized cleaning point.

DAMPERS

A large proportion of the air required for combustion is admitted below the firebars and so is referred to as PRIMARY AIR. This can be regulated by the fireman by means of dampers fixed into the ashpan in the form of swinging doors. There is usually one leading and one trailing damper and are controlled by levers in the cab. The dampers should be operated in such a way as whichever direction the locomotive is travelling the opposite damper door should be used. Whenever the locomotive is about to enter a tunnel the dampers should be fully closed, the blower put on and the firehole doors closed to prevent a blowback.

Boiler Mountings and Boiler Controls

In addition to the boiler being of sound design and construction to be able, to withstand the forces of steam pressure within, certain safety features or safety devices must be. fitted to it to prevent the operating conditions becoming dangerous through over pressure or low water level. Both of these conditions could present a very serious hazard to anyone within a very wide area and in the event of a boiler explosion would cause catastrophic damage to the surrounding area.

On the next page, study the photographs taken in the aftermath of a boiler rupture. These pictures show what incredible forces are stored up within a locomotive boiler, and if allowed to get out of control and the boiler plates fail, anyone on the footplate would be killed instantly or at the very least, very badly scalded. Fortunately these incidents are rare nowadays due to more controlled inspections, however the locomotive boiler is still under full manual control and should be under constant supervision when in steam.

By way of another example to illustrate the enormous power stored up within the locomotive boiler, consider the following simple calculation.

The average locomotive boiler backplate is approx 60in wide and 84in high and the boiler pressure is, say, 2501b in.

60 × 84 = 5040 sq. inches

5040 × 250 = 1,260,000 lbs.

1,260,000 2,240 = 562.5 TONS

This is the steam pressure in front of you when you are stood on the footplate. Just something to consider when you are next booked out on a loco.

Safety Valves

The safety valves are perhaps the single most important safety device fitted to the boiler. They are a mandatory requirement and are fitted to prevent the boiler pressure from exceeding the registered working pressure of the boiler. This is the steam pressure for which it was designed and is indicated by a metal tablet secured to the firebox backplate.

The "Ross Pop" type of safety valve was in extensive use on the former British Railways and it is still in use today. In this design, when the working pressure is reached, the spring loaded valve rises and admits a small amount of steam through the seat of the valve into the Annular Chamber, this gives increased lift to the valve against the pressure of the spring The steam then escapes into the body of the valve and escapes through the holes on the top cap. The steam, when escaping acts on the increased area of the top cap, forcing the valve open still further until such time as the pressure in the boiler has dropped slightly. The spring then overcomes the pressure of the escaping steam and closes the valve instantaneously with a "pop" action.

Blowing off is wasteful of steam and can be avoided by careful management of the fire and injectors. For example: On a large 4-6-0 locomotive, for each minute the safety valve is blowing there is a loss of water of approximately 15 gallons and a loss of fuel of approximately 10 lb.

Water Gauges

Probably the most serious condition any steam boiler can suffer, is low water level. The crown of the firebox must always be covered with water at all times when the boiler is in use. If it should become uncovered, even for a short time, the firebox roof will very quickly become red hot. When this happens the metal it is made of becomes soft and plastic, the steam pressure then forces it off the supporting stays disgorging the remaining steam and water contents into the firebox with violent force. As it is not possible to see through the steel plates, some form of visible water indication is required, and this is performed by the water gauge. The designs consist typically of a top steam cock "K', a bottom water cock 'V' and a drain or blow through cock "C". Cocks "X' and "B" are connected together with a glass tube called the gauge glass. Fitted around the gauge glass is a protector, which is made from thick toughened glass, this protects the glass from being damaged and also protects the train crew in the event of a burst gauge glass.

The water level in the boiler is shown inside the glass tube. With the boiler in steam. The water level must always be visible. With it showing in the bottom nut the firebox will still be covered. The Perforated plate behind the protector usually has on it a set of diagonal black fines, with NO water showing in the glass these black lines will appear the same. With water showing, the part that contains water will be refracted and the black lines will appear to show the other way round. This is one way to tell whether there is water in the glass or not (i.e. Completely full or empty).

Testing Gauge Glass

The gauge glass must be tested at least once after taking over the locomotive and at regular intervals during the time that you have control of the boiler The important thing to remember when testing the gauge glass is that from the top cock "X' the steam should HISS. S. And from the bottom cock "B" the water should ROAR. When it comes out of the drain pipe.

How to change a leaking or burst gauge glass:

- If the gauge glass bursts, throw a coat or sack over the effected fitting, and shut off the steam cock "X' and water cock 'M" instantly
- Open the drain cock "C"
- Remove the perforated backplate and remove the protector
- Remove the top plug from the top fitting and remove the loose restrictor
- Remove the gland nuts from the top and bottom fitting and clean out any remaining glass and rubber washers
- Place the new glass in from the top (if possible) feeding the parts on to the glass as follows:
  - Top rubber washer
  - Top gland follower
  - Top gland nut
  - Bottom gland nut
  - Bottom gland follower
  - Bottom rubber washer
- Push the glass firmly down into the bottom fitting
- Feed in the rubbers, gland followers and finally screw up the gland nuts HAND TIGHT only
- Replace the restrictor and the top plug in the top fitting
- Refit the protector and the perforated back plate, and ensure it is secure
- Warm the glass through by opening the steam cock "X' slowly
- Slowly close the drain cock "C" (the glass may burst again)

- Finally open the water cock 'S" and check the water level rises in the glass.

Boiler Pressure Gauge

In order to detect the steam pressure within the boiler, a pressure gauge is fitted. This is a mandatory requirement and is always fitted directly to the boiler. A cock is usually fitted to allow changing of the gauge should it become defective. The design is always of the Bourdon tube type, it made from phosphor bronze. The dial has a scale and is usually marked off in Pounds per Square Inch. (LBf/in2) though sometimes it may have a dual scale marked in Bars, usually in increments of ten. On every boiler gauge there is marked a Red fine on the scale, this is the maximum working pressure of the boiler and the point at which the safety valves will lift.

Fusible Plugs

In order to give the fireman a warning and to some degree protect the crown of the firebox, One or more fusible plugs are screwed into the roof of the firebox. These are made from bronze and have a Lead core which will melt at quite a low temperature. If the water level drops too low and uncovers the plugs, the lead melts and allows steam to escape into the firebox. The hole the lead fills is only about 318 in. Dia. And so will not be sufficient to extinguish the fire.

Blower Valve and Ring

The blower consists of a perforated ring or pipe fitted round the top of the blast pipe and is connected to a steam supply from the boiler via a valve in the cab and is usually under the control of the driver. The function of the blower is to create a vacuum in the smokebox for the following;

- To increase the draught on the fire when the locomotive is stationary, or in order to raise steam pressure.
- To clear smoke.
- To counteract back draught in the case of 3. Whilst

working a train or light engine, the blower valve must always be opened prior to closing the regulator, or before entering a tunnel.

Blowdown Valve

In order to remove the contaminants of the boiler water which has allowed to settle as sludge at the bottom of the foundation ring, a Blowdown valve is fitted. When this is opened the boiler pressure forces out this sludge with violent but controlled force. Before operating the Blowdown valve the boiler water level must be high in the glass and have sufficient steam pressure to operate the injectors to be able to restore the water level when the valve is closed.

The water level must not be allowed to drop out of sight in the gauge glass, and must be continuously watched when blowing down, with the boiler in steam. The operating handle should be secured when not in use to prevent tampering or inadvertent operation. Care must also be taken when sighting the locomotive for blowing down and a clear warning given to people nearby who may be scalded from the outlet pipe.

REGULATOR VALVE

The regulator valve is probably the biggest boiler fitting and is positioned usually within the dome of the boiler. It is operated by the handle in the cab, mounted on the backplate where it passes through a gland into the boiler. There are several types of valve all designed to provide a controlled passage of steam from the boiler through the superheater (if fitted) to the cylinders. The most common type is the Vertical Slide which has two, three or four ports on the valve (sometimes called Second valve), and two ports on the pilot valve (First valve). The pilot valve opens first when the regulator handle is moved which gives controlled starting. The other types of regulator valve are: Horizontal slide type (fitted to many Ex LMS classes); the Double beat type (fitted to many Ex LNER classes); and the Balanced piston type which are fitted to all the EX SR West Country, Battle of

Britain and Merchant Navy classes. Some types of regulator are fitted in the smokebox within the superheater header.

FIRE TUBE BOILER

Fire-tube boilers are those device in which hot gases, which originates from the fire passes through one or more tubes within the boiler. After three main historical forms of boilers (low-pressure tank or haystack boilers, tube boilers, high-pressure water-tube boilers), fire tube boilers are manufactured. In these boilers, hot gases pass through one or more tubes running through a sealed container of water. According to thermal conduction, the heat energy of the gases pass away from the sides of the tubes, resulting into the heating of water and finally steam.They are important boilers that are available in both horizontal and vertical configurations. They are also some times referred as smoke-tube boiler.

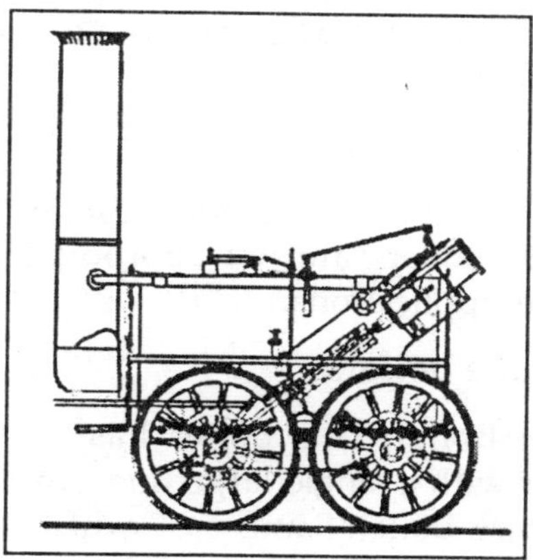

Construction and Usage

Usually, the tank of the boiler is cylindrical in shape. It can also be either vertical or horizontal. Also regarded as shell boiler, smoke tube boiler and fire pipe, fire tube boilers are highly used on all steam locomotives in the horizontal locomotive form. Other than used as locomotives, they are extensively used in stationary engineering fields and producing low pressure steam like for heating of a building.

Operation

The fuel is burnt in a firebox to generate hot combustion gases. Steam is produced by heating water along a series of fire tubes perforating the boiler. The generated steam raises the steam dome, which also controls the exit of steam. In regards to superheat and dry generated steam, saturated steam is always forced into the super heater through the larger flues at the back in the locomotive fire tube boiler. With the help of steam directed to the turbine or cylinders, mechanical power is induced and the exhaust gases are used to heat the feed water for increasing boiler's efficiency.

Advantages

Fire tube boilers are perfectly used for space heating and various industrial process applications. They are compact in size, easy to clean and maintain. The tubes are easily replaceable and cost maintenance is very low. They are available in wide range of sizes.

Disadvantages

- They are limited for generating high capacity steam.
- Fire tube boilers are not appropriate for very high pressure applications.

Types of Fire tube boilers

- Cornish Boiler: This boiler has a long horizontal cylinder with a single large flue consisting the fire.
- Lancashire boiler: Same as Cornish boiler, this has two large flues comprising the fires.
- Scotch marine boiler: This form of boiler uses large number of small diametre tubes. It provides greater heating surface arena for both weight and volume.
- Locomotive boiler: It has three major constituents, double walled firebox, horizontal and cylindrical boiler barrel consists with small flue tubes and smoke box provided with chimney for exhaust gases. They are vastly used in traction engines, steam rollers, portable engines and some other steam road vehicles.

BOILER MOUNTINGS AND ACCESSORIES

BOILER MOUNTINGS

Being a decade old distributor of WJ Boiler Mountings, we have become the most preferred choice of the market. With the WJ boiler mountings at hand we have an edge over every other steam boiler mountings suppliers of India. The safety boiler mountings that we deal in are of the best quality; leaving behind every speculation about us and of the products. These products are assured to have high durability, thus, proving to be highly advantageous in the long run. These distinct boiler mountings are available in the market at some of the most pocket friendly prices.

It includes:

- Feed check valve
- Pressure reducing valve
- Steam stop valve
- Safety valve
- Strainer
- Accessible feed check valve
- Blow off cock
- Non-return valve
- Steam trap
- Manifold

Safety Flow Control Valves

We are mainly engaged in supplying of safety flow

control valves, which includes gate, globe, check, wafer type single/dual plate check, ball, butterfly, diaphragm, self lubricated plug type, knife edge, jacketed sluice, penstock, needle, non-return, flush bottom, balancing, breather etc in different material of construction.

IMPORTANT BOILER MOUNTINGS

A boiler just cant run on its basic working system. There are many external equipments that help a boiler run efficiently. For example, it would have been too difficult to keep a continuous watch on the boiler water level had there not been cascade or boiler automation system. In the same way there are a few important boiler mountings that are fitted on the boiler body itself for a safe and efficient running of the boiler. These fittings are the basic requirements of the whole steam production system. The main boiler mountings are:

Safety Valves

You would always find safety valves mounted in pairs on main boiler body. Safety valves are provided on a boiler to protect the boiler from overpressure. All the safety valves are preset at a particular fixed boiler pressure. This means that the valves are set at a particular lifting pressure and locked to prevent any kind of manipulation later on. The safety valves operate automatically when the boiler reaches the pre-set blow off pressure.

Main Steam Valve

This is the most important valve of the whole system. Main steam stop valve is a non return valve, fitted on the main steam line just near to the boiler. It is the main valve by which the supply of steam from the boiler to all the ship's systems can be started or stopped.

Auxiliary Stem Stop Valve

This is like a back up valve to the main steam stop valve. This means that in case some problem occurs with the main steam stop valve, the auxiliary valve still protects the system from any kind of damage. It is also a non return valve type.

Feed Check or Control Valve

These valves are also found in pairs. One of the valve out of the two valves is the main valve and the other is an auxiliary or stand by valve. These valves are also non return valves and they always carry an indication of their open or closed position.

Feed Check or Control Valve

These valves are also found in pairs. One of the valve out of the two valves is the main valve and the other is an auxiliary or stand by valve. These valves are also non return valves and they always carry an indication of their open or closed position.

Water Level Gauges

Water level gauges are one of the most important mountings of the boiler. Always found in pair, the water level gauges help in maintaining and monitoring the appropriate water level in the boiler. Thus they help in preventing accidents and damage to the boiler. The gauges are always fitted at the opposite ends of the boiler to prevent wrong reading due to motion of the ship. The type of the water gauge fitted depends on the the working pressure of the boiler

Pressure Gauge Connection

Pressure gauges are fitted on the boiler drums and superheaters for pressure readings. They should be periodically checked to prevent accidents due to wrong readings.

Air Release Cock

An important equipment for the boiler starting procedure, the air release cock are fitted in boiler drums, headers etc to purge out the air when filling the boiler or during raising steam

Sampling Connection

This connection is mainly for taking feed water samples for testing and analysis purposes. The samples are collected through this connection and if any kind of chemicals that are to be added to the boiler water they are injected through the this connection.

Blow Down Valve

Used for emptying the boiler water, the blow down valve is fitted at the lowermost end of the boiler. It is also used for partial emptying the boiler and is an important valve in the blow down process.

Scum Valve

This valve is used for removal of impurities from the boiler water. The scum or impurities float on the top of the water. A shallow dish is positioned at the normal water level to collect the scum from the top of the water level.

Whistle Stop valve

It is small bore non return valve which is used to supply steam to the whistle straight from the boiler drum

Additional Mountings on Water Tube boiler

Water tube boilers use less water and produce more steam in comparison to it.Due to this reason the chances of accidents due to high temperature and pressure are greater. Due to this reason, water tube boilers have extra mountings

Automatic Feed Water Regulator

Fitted in the feed water line just near the main check valve, the automatic feed water regulator ensures correct water level in the boiler in spite of the fluctuating load conditions. Boilers having high evaporation rate use multiple element feed water regulator

Low Alarm Level

Produces an alarm sounding when the boiler reaches low water level condition.

High Alarm Level

Produces an alarm sounding when the boiler reaches high water level condition

Super Heater Circulating Valves

Used during the initial warming up process and during the boiler starting procedure, the valves ensures a steady flow of steam and also act as air vents

Sootblowers

Generally used during the boiler maintenance procedure.Sootblowers use pressurized steam or compressed air to blow off the soot and other combustion products deposits from the surface of the tube.

Soot Blow Valves Work On Steam Use To Clean Surfaces In Water Tube Boiler From Deposit Of Fuel Dust Otherwise Rate Of Heat

Transfer Reduces and Boiler Steam Generation and Efficiency Also Reduces.

Types:

- R otary Valve (Manual and Motorised)
- Retractable Long and Short Type.

Gate/Globe Valve, High Pressure Gate/Globe Valves As Up To 63 Kg/cm2. gauge glass assembly.

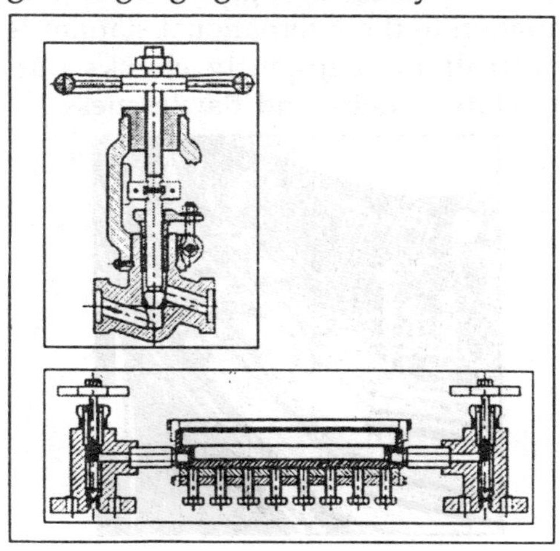

BOILER ACCESSORIES

Boiler Integral Piping

Our team of adept professionals intricately designs our Boiler Integral Pipings. They utilize various sophisticated machinery and innovative techniques in the fabrication of our range. In our endeavor to satisfy our clients all across the globe, we also manufacture our gamut in customised form adhering to the designs and specifications of our esteemed clients.

Turbine Integrated Piping

We are available with Turbine Integrated Piping that finds wide application indifferent industries worldwide. These are fabricated using best grades of raw material and are in conformation to the international standards. Our team of quality controllers stringently checks our range for precision, durability, quality and flawlessness.

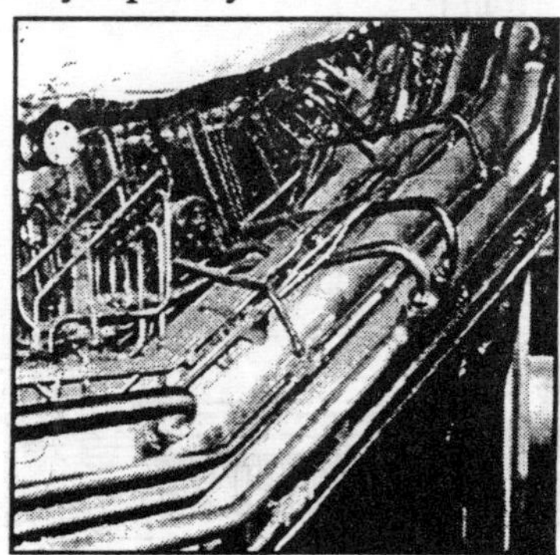

HRSG Integrated Piping

The HRSG Integrated Piping offered by us are of world class designs and are in high demand in the domestic market as well s the foreign front. These are high in performance, low on maintenance and are durable, corrosion resistant and dimensionally accurate.

Risers and Down Comers for Boilers

Over the years we have been successfully catering our broad client base spread all across the globe offering an unmatched quality range of Risers and Down Comers For Boilers. These are intricately designed using sophisticated technologies and hi-tech machines and conform to the international standards.

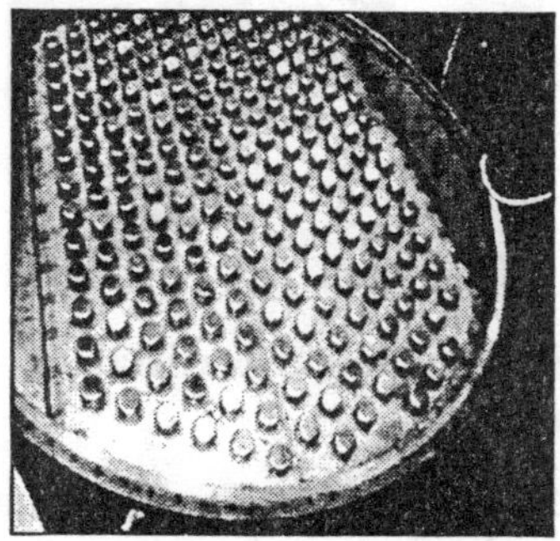

Balance of Plant Piping

The Balance Of Plant Pipings offered by us are in high demand worldwide. These are meticulously designed and breathe in quality, durability, and flawlessness. With our unparallel gamut, we have fruitfully attained the market credibility.

Cooling Water and Lubrication Piping

We expertise in the smooth fabrication of Cooling Water and Lubrication Piping that are high in performance, low on maintenance, durable and sturdy. We offer our range in customised form adhering to the designs and specifications of our esteemed clients so as to provide maximum satisfaction to them.

Stainless Steel Piping for Desalination and Process Plants

Our Stainless Steel Piping For Desalination and Process Plants are exclusively designed and are in high demand in different industries worldwide. These are fabricated in conformation to the international standards and are stringently tested as per well-defined parametres.

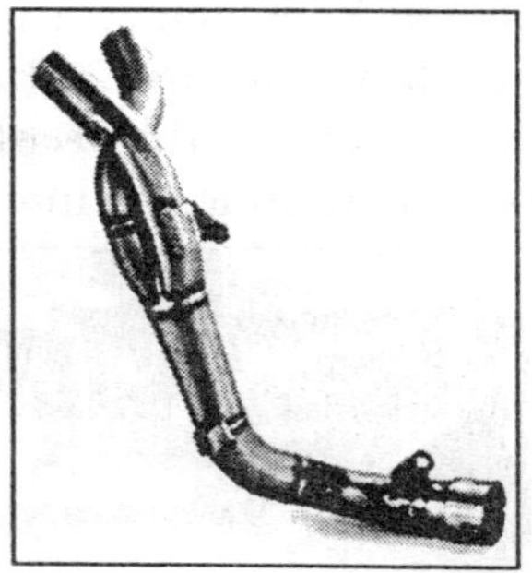

Process Skids

The Process Skids manufactured by us bear world-class designs and are fabricated using best grades of raw material procured from reliable sources. These are highly efficient and are also available in customised form. Besides we also package these equipments as per the requirements of our esteemed clients.

BOILER COMPONENTS

Our cutting edge technology backed by a pool of experienced professionals aid in the smooth fabrication of our wide assortment of Boiler Components including boiler drums, boiler super heaters, economizer coils, boiler headers, boiler r&w drums, industrial boiler drums, fabricated boiler drums, heaters, super heaters, industrial boiler super heaters, fabricated boiler super heaters, cooling coils, heating coils.

Boiler Headers (Steam Distribution Headers)

The Boiler Headers are fabricated using premium quality

raw material and are in high demand in the market. These are also offered in customised form adhering to the designs and specifications of our valued customers and are manufactured in conformation to the international standards.

Boiler Super Heaters

Our Boiler Super Heaters are designed to perfection and are stringently tested as per well-defined parametres. These are meticulously designed by our qualified team of professionals and are widely used all round the globe in different industries and are stringently checked for quality, durability and flawlessness.

Boiler R&W Drums

The Boiler R&W Drums available with us are in high demand all across the globe. Our team of qualified personnel take every care in the production of the best range ever and delivering them in an estimated time frame. Further, we endeavor to satisfy our clients by manufacturing our range as per their requirements.

Economizer Coils

We are engaged in the manufacturing and exporting Economizer Coils that are high in performance and low on maintenance. Designed to perfection these are widely used in different industries all across the world and are fabricated from premium quality raw material conforming to the international standards.

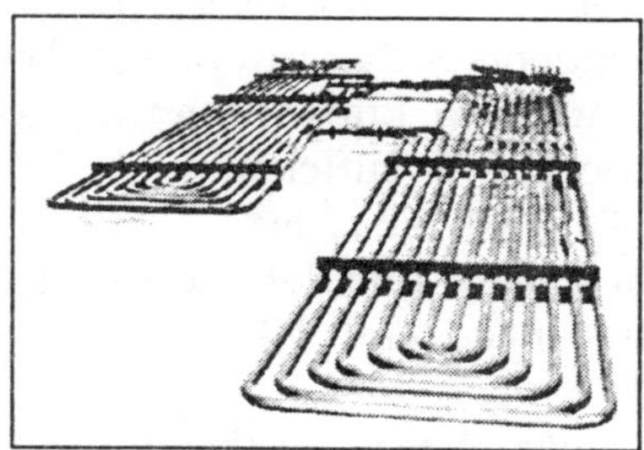

PROPERTIES OF WORKING FLUID FOR POWER PLANTS

DESIRABLE PROPERTIES

Properties of Decomposition

1. Another desirable property in database design is dependency preservation.

- We would like to check easily that updates to the database do not result in illegal relations being created.
- It would be nice if our design allowed us to check updates without having to compute natural joins.
- To know whether joins must be computed, we need to determine what functional dependencies may be
  tested by checking each relation individually.
- Let F be a set of functional dependencies on schema R.
- Let {R1,R2...,Rn} be a decomposition of R.
- The restriction of F to Ri is the set of all functional dependencies in F+ that include only attributes of Ri.
- Functional dependencies in a restriction can be tested in one relation, as they involve attributes in one
  relation schema.
- The set of restrictions F1, F2..., Fn is the set of dependencies that can be checked eciently.
- We need to know whether testing only the restrictions is suffcient.
- Let F0 = F1, F2..., Fn.
- F0 is a set of functional dependencies on schema R, but in general, F0
  6= F.
- However, it may be that F'+ = F+.
- If this is so, then every functional dependency in F is implied by F0, and if F0 is satis ed, then F must also be satisfied.
- A decomposition having the property that F0+ = F+ is a dependency-preserving decomposition.

2. The algorithm for testing dependency preservation follows this method.

compute F+ for each schema Ri in D do
begin

Fi.= the restriction of F+ to Ri,
end
F′ = Ð¤,
for each restriction Fi do
begin
F′ = F Fi
end
compute F′+,
if (F′+ = F+) then return (true)
else return (false),

3. We can now show that our decomposition of Lending-schema is dependency preserving.
   - The functional Dependency bname → assets bcity can be tested in one relation on Branch-schema.
   - The functional Dependency loan# → amount bname can be tested in Loan-schema.
4. As the above example shows, it is often easier not to apply the algorithm shown to test dependency preservation, as computing F+ takes exponential time.
5. An Easier Way To Test For Dependency Preservation. Really we only need to know whether the functional dependencies in F and not in F0 are implied by those in F′. In other words, are the functional dependencies not easily checkable logically implied by those that are?
   Rather than compute F+ and F′+, and see whether they are equal, we can do this.
   - Find F - F′, the functional dependencies not checkable in one relation.
   - This should take a great deal less work, as we have (usually) just a few functional dependencies to work on.

Desirable Properties of a Database

As you will see, there are many possible choices that can be made during the design and many rules to guide this work. When trying to decide if some choices are better than others, you need to consider the key desirable properties of a database. The table here outlines some of them:

Completeness	Ensures that users can access the data they want. Note that this includes **ad hoc queries**, which would not be explicitly given as part of a statement of data requirements.
Integrity	Ensures that data is both consistent (no contradictory data) and correct (no invalid data), and ensures that users trust the database.
Flexibility	Ensures that a database can evolve (without requiring excessive effort) to satisfy changing user requirements.
Efficiency	Ensures that users do not have unduly long response times when accessing data.
Usability (ease of use)	Ensures that data can be accessed and manipulated in ways which match user requirements.

Developing a 'good' database with these desirable properties isn't easy. Indeed, it is quite possible to develop a database that appears to contain all the relevant data but does not have these properties: as you will see, this has unfortunate consequences.

Exercise: Using the given desirable database properties above, very briefly summarise the characteristics that may be expected with a 'bad' database.

The importance of good database development is found simply in terms of preventing the problems outlined in Solution 2. Database development is not just a matter of creating tables that seem to match the way you see data on forms or reports, but requires a detailed understanding of the meaning of the data and their relationships to ensure that a database has the right properties. This understanding comes from data analysis, which is concerned with representing the meaning of data as a conceptual data model. Getting a conceptual data model right is crucial to the development of a good database.

Desirable Properties of Real-time

Each of these categories is expanded upon below, and later used to compare a number of proposed realtime approaches for Linux. The discussion does go for some time, which is not surprising given that it is summarizing many hundreds of email messages:

- Quality of Service

The traditional view is that the entire operating system is either hard realtime, soft realtime, or non-realtime, but this viewpoint is too coarse grained. Different workloads have different needs, and there is disagreement over the exact definitions of these three categories of realtime. For example, (at least) the following two definitions of "hard realtime" are in use:

- In absence of hardware failures, software provably meets the specified deadlines. This is fine and good, but many applications simply do not need this "diamond hard" realtime.
- Failure to meet the specified deadline results in application failure. This is OK, but -only- if there is a corresponding required probability of success. Otherwise, one could claim "hard realtime" by simply failing the application every time it tries to do anything, which is clearly not useful.

A better approach is to simply specified the required probability of meeting the specified deadline in absence of hardware failure. A probability of 1.0 is consistent with definition (a). Other applications will be satisfied with a probability such as 0.999999, which might be sufficiently high that the probability of software scheduling failure is "in the noise" compared with the probability of hardware failure. A recent LKML thread called this "metal hard" realtime. Or was it "ruby hard"?;-) Of course, one can increase the reliability of hardware through redundancy, but no hardware configuration provides perfect reliability. For example, clusters can increase reliability, so that the probability of failure of the cluster is p^n, where "p" is the probability of a single node failing and "n" is the number of nodes. Note that this expression never reaches a probability of 1, no matter how large "n" is. In addition, this mathematical expression assumes that the failover software is perfectly reliable and perfectly configured. This assumption conflicts sharply with my own experience, in which there has always been a point beyond which adding nodes -decreased- cluster reliability.

The timeframe is also critically important. Any system can provide hard realtime guarantees if the deadline is an infinite amount of time in the future. No computer system that I am aware of at this writing is capable of meeting a 1-picosecond scheduling deadline for any task of non-zero duration, but then neither can dedicated digital hardware. Some applications have definite response-time goals, for example, industrial process-control applications tend to have response-time goals ranging from 100s of microseconds to small numbers of seconds. Other applications can benefit from any improvement in response-time goals — faster is better, think in terms of Doom players — but even in these cases there is normally a point of diminishing returns.

The services used by the realtime application also figure in. Given current disk technology, it is not possible to meet a 100-microsecond deadline for a 1MB synchronous write to disk. Not even if you cheat and supply the disk with a battery-backed-up DRAM. However, many realtime applications need only a few of the services that an operating system might provide. This list might include interrupt handling, process scheduling, disk I/O, network I/O, process creation/ destruction, VM operations, and so on. Keep in mind that many popular RTOSes provide very little in the way of services! They frequently leave the complex stuff (e.g., web serving) to general-purpose operating systems.

Note that each service can have an associated deadline that it can meet. The interrupt system might be able to meet a 1-microsecond deadline, the real-time process scheduler a 10-microsecond deadline, the disk I/O system a 10-millisecond deadline for moderate-sized I/Os, and so on. The deadline that a service can meet might also depend on the parametres, so that the disk-I/O system would be expected to take longer for larger I/Os.

Furthermore, the probability might vary from service to service or with the parametres to that service. For example, the probability of network I/O completing successfully in minimal time might well be a function of the number of packets transmitted (to account for the probability of packet loss) as well as of packet size (to account for bit-error rate).

To make things even more complicated, the probability of meeting the deadline will vary depending on the length of time allowed. Considering the networking example, a very short deadline might not allow the data transmission to complete, even if it proceeds at wire speed. A longer deadline might allow transmission to complete, but only if there are no transmission errors. An even longer deadline might allow time for a limited number of retransmissions, in order to recover from packet loss due to transmission errors. Of course, a deadline infinitely far into the future would allow guaranteed completion, but I for one am not that patient.

Finally, the performance and scalability of both realtime and non-realtime applications running on the system can be important. Given the current state of the art, one must pay a performance penalty for realtime support, but the smaller the penalty, the better.

So, to sum up, here are the components of a quality-of-service metric for realtime OSes:

- List of services for which realtime response is supported.
- For each service:
  - Probability of missing a deadline due to software, ranging from 0 to 1, with the value of 1 corresponding to the hardest possible hard realtime.
  - Allowable deadline, measured from the time that

the request is initiated to the time by which the response must be received.

  – Performance and scalability provided to both realtime and non-realtime applications.

- Amount of Code Inspection Required

So you add a new feature to a realtime operating system. How much of the rest of the system must you inspect and understand in order to be able to guarantee that your new feature provides the required level of realtime response? The smaller this amount of code, the easier it is to add new features and fix bugs, and the greater the number of people who will be able to contribute to the project. In addition, the smaller the amount of such code, the smaller the probability that some well-intentioned bug fix will break realtime response.

Each of the following categories of code might need to be inspected:

- The low-level interrupt-handing code.
  – The realtime process scheduler.
  – Any code that disables interrupts.
  – Any code that disables preemption.
  – Any code that holds a lock, mutex, semaphore, or other resource that is needed by the code implementing your new feature.

Of course, use of automated tools could make such inspection much more reliable and less onerous, but such tools would need to deal with the very large number of CPU architectures and configuration options that Linux supports. The smaller the amount of code that must be inspected, the less chance there is that such a tool will fall victim to configuration-architecture combinatorial explosion.

Each of Linux realtime approaches uses a different strategy to minimize the amount of code in these categories. These differences are surprisingly important, and will be discussed in more detail when going over the various approaches to Linux realtime.

- *API Provided*: I never have learned to -really- like the POSIX API, with the gets() primitive being a particular cause of heartburn, but given the huge amount of software out there that relies on it and the equally huge number of developers who are familiar with it, one should certainly strive to provide it, or at least a sizeable subset of it.

Other popular APIs include the various Java runtime environments, and of course the feared and loathed, but quite ubiquitous, Windows API.

There are a lot of developers and a lot of software out there. The more of these existing developers and software your API supports, the more successful your realtime facility is likely to be.

- *Relative Complexity*: How much realtime capability should be added to the operating system? How much of this burden should the applications take on? Is it better to push some of the complexity into a nanokernel, hypervisor, or other software or firmware layer? Let's first look at the tradeoff between OS and application.

For example, although it is certainly possible to program for separate realtime and non-realtime operating-system instances, doing so adds complexity to the application. Complexity is particularly deadly in the hard realtime arena, and can be literally so if human lives are at risk.

Balancing this consideration is the need for simplicity in the operating-system kernel. This balancing act must be carefully considered, taking both the relative complexities and the number of uses into account. Some would argue that it is worthwhile adding 1,000 lines to the OS if that saves 100 lines in each of 1,000 applications. Others would disagree, perhaps citing the greater fault isolation that might be provided by the separation. But this balance clearly must be struck somewhere between writing the application to bare metal on the one hand (but achieving a perfectly simple zero-size operating system) and bloating the operating system beyond the limits of maintainability on the other hand.

Similar arguments can be made for moving some functionality into a hypervisor or nanokernel layer, though fault isolation also comes into play here. Many of the most vociferous arguments seem to revolve around this complexity issue.

- *Fault Isolation*: Can a programming error in a non-realtime application or in a non-realtime portion of the OS harm a realtime application?

Some applications do not care: in these cases, a failure anywhere causes a user-visible failure, so it is not important to isolate faults. Of course, even in these cases, it may be valuable to isolate faults in order to aid debugging, but, other than that, the fault isolation does not help overall application reliability.

In other cases, the realtime portion of the application is protecting someone's life and limb, but the non-realtime portion is only compiling statistics and reports. In this case, fault isolation can be of the utmost importance.

What sorts of faults need isolating?

- Excessive disabling of interrupts
- Excessive disabling of preemption
- Holding a lock, mutex, or semaphore for too long, when that resource must be acquired by realtime code
- Memory corruption, either via wild pointers or via wild DMA

These faults might occur in the main kernel, in a loadable module, or in some debugging tool, such as a kprobe procedure or a kernel-debugger breakpoint script. Though in the latter case, perhaps realtime deadlines should not be guaranteed when actively debugging. After all, straightforward debugging techniques, such as use of kprint(), can cause response-time problems even in non-realtime environments.

- Hardware and Software Configurations

Is SMP required? If so, how many CPUs? How many tasks? How many disks? How many HBAs?

If all the code in the kernel were O(1), it might not matter, but the Linux kernel has not yet reached this goal. Therefore, some applications may choose to restrict the software or the hardware configuration of the platform in order to meet the realtime deadlines. This approach is consistent with traditional RTOS methodology – RTOS vendors have been known to restrict the configurations in which they will support hard realtime guarantees.

WORKING FLUID

MERCURY

Mercury has got to be the ultimate dodgy working fluid mercury: as remarked elsewhere on this site, in the Steamwheel gallery, (where mercury is used as a weight and sealant rather than a working fluid) it is an insidious poison of a most unpleasant kind. It has a much higher boiling point than water, at 357 degC. Experimental power station installations were tested in the USA but ultimately nothing came of them. Probably a good thing.

ERCURY IN THE NINETEENTH CENTURY

There was another kind of marine engine that I think should not be passed over without notice; I allude to Howard's quicksilver engine. The experiments with this engine were persevered in for some considerable time, and it was actually used for practical purposes in propelling a passenger steam-vessel called the Vesta, and running between London and Ramsgate. In that engine the boiler had a double bottom, containing an amalgam of quicksilver and lead.

This amalgam served as a reservoir of heat, which it took up from the fire below the double-bottom, and gave forth at intervals to the water above it. There was no water in the boiler, in the ordinary sense of the term, but when steam was wanted to start the engine, a small quantity of water was injected by means of a hand-pump, and after the engine was started, there was pumped by it into the boiler, at each half revolution, as much water as would make the steam needed.

This water was flashed on the top surface of the reservoir in which the amalgam was confined, and was entirely turned into steam, the object of the engineers in charge being to send in so much water as would just generate the steam, but so as not to leave any water in the boiler. The engines of the Vesta were made by Mr Penn, for Mr Howard, of the King and Queen Ironworks, Rotherhithe. Mr Howard was, I fear, a considerable loser by his meritorious efforts to improve the steam-engine."

"There was used, with this engine, an almost unknown mode of obtaining fresh water for the boiler. Fresh water, it will be seen was a necessity in this mode of evaporation. The presence of salt, or of any other impurity, when the whole of the water was flashed into steam, must have caused a deposit on the top of the amalgam chamber at each operation.

Fresh water, therefore, was needed; the problem arose how to get it; and that problem was solved, not by the use of surface condensation, but by the employment of reinjection, that is to say, the water delivered from the hot well was passed into pipes external to the vessel; after traversing them, it came back into the injection tank sufficiently cooled to be used again. The boilers were worked by coke fires, urged by a fan blast in their ashpits, but I am not aware that this mode of firing was a needful part of the system."

The idea of using an amalgam of mercury and lead as a reservoir of heat was not a good one, and one regrets that Mr Howard was a considerable loser. The liquid with the highest specific heat, and so greatest storage capacity, is not some exotic chemical. Remarkably, it is plain water. Much cheaper and much safer!

MERCURY IN THE TWENTIETH CENTURY

The man behind the use of mercury vapour turbines in electric power stations was William Le Roy Emmet of the General Electric Company. Emmet devoted a great deal of time and energy to their development and promotion, "as a more efficient power generation system than steam turbines." The difficulties and risks in using the new technology led to termination of the development after his retirement. He died in Erie, Pennsylvania, on September 26, 1941, at the age of

82, so he had presumably managed to avoid mercury poisoning.

Mercury has a much higher boiling point than water, so the Carnot efficiency of a thermodynamic cycle that uses it is higher, in theory at least. The plants were dual-cycle, (also known as binary cycle) the hot metallic exhaust from the mercury turbines being used to produce steam for a second stage of steam turbines.

The mercury is vapourised in the boiler, drives the mercury turbine, and is condensed by boiling water to steam. This steam is superheated in the mercury boiler and then drives a conventional steam turbine. The "bled steam heaters" refer to the practice of heating the feed water by bleeding steam from various points along the turbine; this increases cycle efficiency and had been a standard feature of steam power plants for many years.

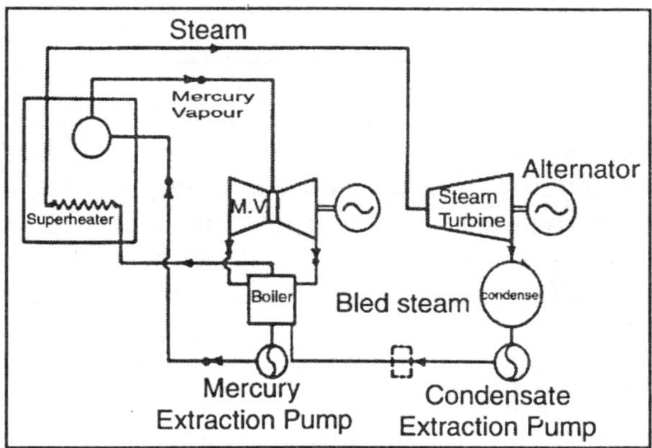

On close inspection this diagram shows exactly the same plumbing as the one above.

According to this book, the mercury circuit operated at 200psi and 538 degC. Electrically welded joints were used in the pipework because the poisonous nature of mercury was recognised, and leaks were not acceptable. Even so, it is hard to believe that a big installation could have been completely free from leaks, and I for one would not have cared to work there, or indeed within ten miles of it.

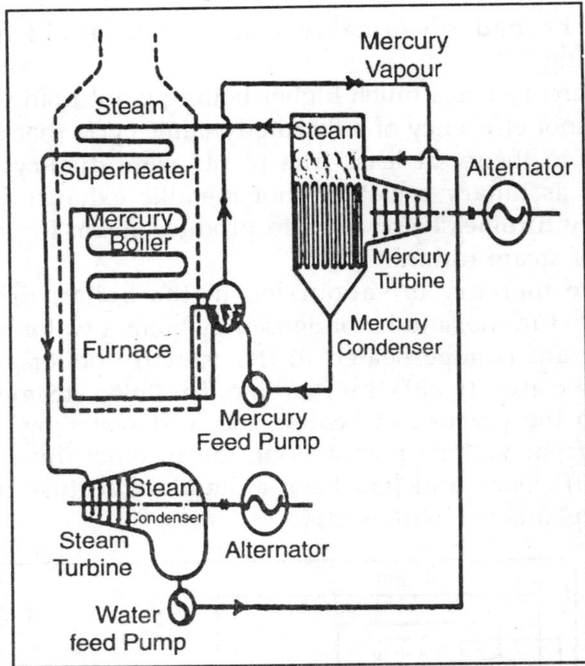

In one American plant (it is not clear which) the mercury circuit operated at 113 psi and 507 degC, driving a 15 MW turbine-alternator set at 720 rpm. The mercury condenser produced steam at 400 psi and 371 degC, at the rate of 200,000 lb/hour; this drove conventional steam turbines and alternators.

Efficient Use of Steam

"The Efficient Use of Steam" by Oliver Lyle was the bible of steam usage in Britain during the second world war and after. It was aimed more towards process steam rather than power generation, on mercury/water binary cycles: "Mercury boils at 900F under the very modest pressure of 80psi. At 28.5 in vacuum mercury vapour has a temperature of 437 degF (225 degC).

If therefore, mercury is boiled in a boiler, and the saturated mercury vapour passed through a mercury turbine the mercury vapour can be exhausted into a condenser which can be a water boiler, and can raise steam to 250psi.This steam can be given some superheat up to say 600 degF (316 deg C) from the flue gases of the mercury boiler. Owing to the low latent heat of mercury it is necessary to use about 10 pounds of mercury for every 1 pound of water." and later in the text:

"The disadvantages are somewhat formidable. Mercury vapour is extremely poisonous. Mercury does not wet metal surfaces.* The plant is complicated and costly. But several large mercury-steam stations have been working for some years in the USA and have shown very high sustained overall thermal efficiencies. other fluids might give better results. Diphenyl oxide has been suggested as being more suitable than mercury."

If the surfaces are not wetted by the working fluid, then heat transfer would be greatly impaired. But you will see below that mercury *does* wet tantalum. Now that's an interesting point above about the amount of mercury required. And it's not cheap stuff.

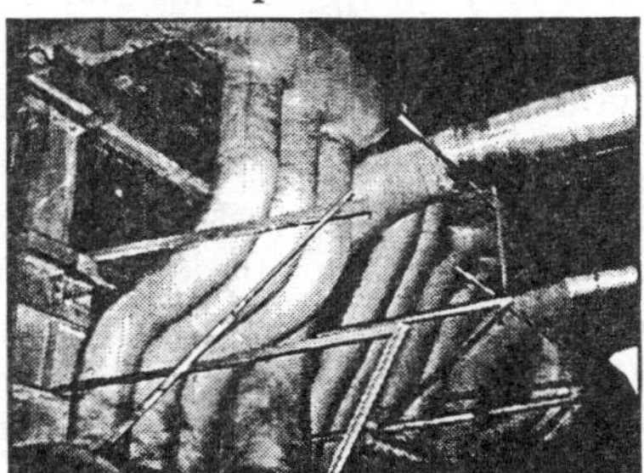

Mercury Power in Space

The reactor is cooled with a mixture of sodium and potassium, (NaK) which heats a mercury boiler. The mercury vapour drives a turbo-alternator and is then condensed and subcooled by a secondary NaK heat rejection loop which transfers the waste heat to a radiator for rejection to space.

The report concentrates on the design of the potassium/ mercury boiler, which is an exotic device indeed. It is a spiral stucture designed to fit in a small toroidal space in a cylindrical spacecraft; the heat exchanger tubes are of tantalum, which is readily wetted by mercury, making for good heat transfer. The net power output was 37 kW.

Note the extra cooling and lubricating loop using polyphenyl ether.

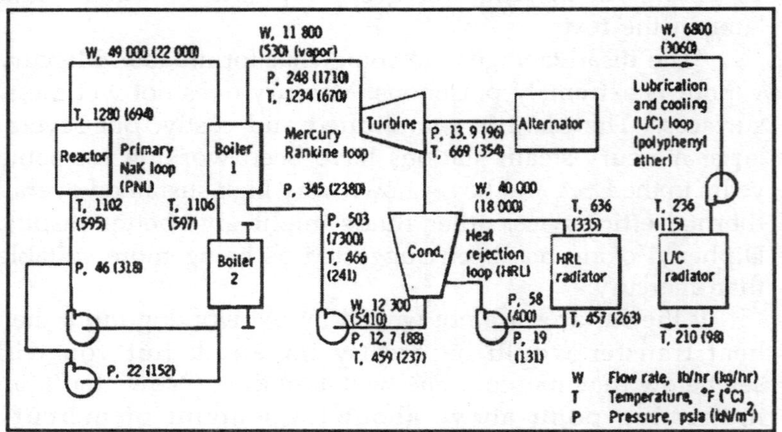

The extraordinary thing is that the overall efficiency seems horribly low at 6.9%. A conventional terrestrial steam power station would give some where around 40%.

So why is it so bad? The Carnot efficiency E = (T1 - T2)/ T1 where T1 is the heat input (top) temperature and T2 the heat rejection temperature. The effective rejection temperature in the condenser is a bit less clear but the NaK coolant leaves it at 335 degC. This gives a Carnot efficiency of 35%. That's obviously a theoretical maximum, but a long step from 6.9%; there must be some serious losses somewhere.

The rejection temperature has to be high because in space there are no rivers flowing by to cool your condensers- the only way to lose heat is to radiate it away from a flat plate, which needs to be pretty hot to radiate effectively. Hence the radiators in the flow diagram; these are nothing like central-heating "radiators" which work mostly by convection.

This is one reason- possibly the most important- for the choice of mercury as a working fluid; it permits a high heat rejection temperature so the radiators are efficient, and therefore relatively small and light.

This project was only one of several nuclear-powered sytems that were intended to provide large quantities of electricity for long-duration space missions. One of the reasons that they have not so far been used is that no-one is very comfortable with the notion of launching a nuclear reactor on top of a rocket.

USED FOR POWER PLANT

Three types of low toxicity gas-oil burners have been developed at the firm JSC "NPO TsKTI." Thus far, they have been introduced at a number of electric power stations in this country for converting coal dust burners to combustion of natural gas and oil and for modernizing gas-oil burners, making use of well known in-boiler measures for suppressing the formation of nitrogen oxides. In this way the engineering-economic and ecological performance of the boilers has been improved significantly.

Our renewable energy engineering and renewable energy project development services include: Carbon Credits and Carbon Emissions Consulting, Design, Engineering, Environmental, Feasibility Studies, Feedstock, Legal, Onsite Power Generation (cogeneration or trigeneration) & Greenhouse Gas Emissions consulting for projects located in the U.S. and Canada.

A Circulating Fluidized Bed Boiler is a fully contained state-of-the-art technology for processing solid fuels where fuel is suspended in a mixture of superheated air and sand, collectively called the "fluid bed." Reagents like limestone are added, and temperatures are controlled to directly capture the sulfur and reduce formation of Nitrogen Oxides.

Fluidized Bed Boilers, a Bed for Burning Coal

It was a wet, chilly day in Washington DC in 1979 when a few scientists and engineers joined with government and college officials on the campus of Georgetown University to celebrate the completion of one of the world's most advanced coal combustors. It was a small coal burner by today's standards, but large enough to provide heat and steam for much of the university campus. But the new boiler built beside the campus tennis courts was unlike most other boilers in the world.

A Fluidized Bed Boiler

In a fluidized bed boiler, upward blowing jets of air suspend burning coal, allowing it to mix with limestone that absorbs sulfur pollutants.

It was called a "fluidized bed boiler."

In a typical pulverized coal boiler, coal is crushed into very fine particles, blown into the boiler, and ignited to form a long, lazy flame. In other types of boilers, the burning coal simply rests on grates. But in a "fluidized bed boiler," crushed coal particles "float" inside the boiler, suspended on upward-blowing jets of air. The red-hot mass of floating coal — called the "bed" — would bubble and tumble around like boiling lava inside a volcano. Scientists call this being "fluidized." That's how the name "fluidized bed boiler" came about.

Fluidized Bed Boiler asBurn Coal Cleaner

There are two major reasons fluidized bed boilers are cleaner, and superior to typical coal fired power plants. One, the tumbling action allows limestone to be mixed in with the coal. Remember - limestone is a "sulfur sponge" in that it absorbs sulfur pollutants. As coal burns in a fluidized bed boiler, it releases sulfur. But just as rapidly, the limestone tumbling around beside the coal captures the sulfur. A chemical reaction occurs, and the sulfur gases are changed into a dry powder that can be removed from the boiler.

The second reason a fluidized bed boiler burns cleaner is that it burns "cooler." Cooler in this sense as it is still fairly hot at about 1400 degrees F. But older coal boilers operate at temperatures nearly twice that (almost 3000 degrees F). Also, recall that nitrogen oxides form when a fuel burns hot enough to break apart the nitrogen molecules in the air and cause the nitrogen atoms to join with oxygen atoms. But 1,400 degrees isn't hot enough for that to happen, so few nitrogen oxides forms in a fluidized bed boiler. The result is that a fluidized bed boiler can burn very dirty coal and remove 90% or more of the sulfur and nitrogen pollutants while the coal is burning. Fluidized bed boilers can also burn just about anything else - all types of biomass, including wood, ground-up railroad ties, even soggy coffee grounds.

Today, fluidized bed boilers are operating or being built that are 10 to 20 times larger than the small unit built almost 20 years ago at Georgetown University. There are more than 300 of these boilers operating here in the USA and around the world.

A new type of fluidized bed boiler makes a major improvement in the basic fluidized bed boiler technology. It encases the entire boiler inside a large pressure vessel, much like the pressure cooker used in homes for canning fruits and vegetables — except the ones used in power plants are the size of a small house! Burning coal in a "pressurized fluidized bed boiler" produces a high-pressure stream of combustion gases that can spin a gas turbine to make electricity, then boil water for a steam turbine — two sources of electricity from the same fuel input - that is called "cogeneration."

A "pressurized fluidized bed boiler" is a more efficient way to burn coal. In fact, future boilers using this system will be able to generate 50% more electricity from coal than a regular power plant from the same amount of coal. That's like getting 3 units of power when you used to get only 2. Because it uses less fuel to produce the same amount of power, a more efficient "pressurized fluidized bed boiler" will reduce the amount of carbon dioxide (a greenhouse gas) released from coal-burning power plants.

Coal Gasification

One of the most advanced - and cleanest - coal power plants in the world is Tampa Electric's Polk Power Station in Florida. Rather than burning coal, it turns coal into a gas that can be cleaned of almost all pollutants. This technology is called coal gasification. How do you break apart the atoms of coal? You may think it would take a sledgehammer, but actually all it takes is water and heat. Heat coal hot enough inside a big metal vessel, blast it with steam (the water), and it breaks apart. Into what?

The carbon atoms join with oxygen that is in the air (or pure oxygen can be injected into the vessel). The hydrogen atoms join with each other. A major reason is that the impurities in coal — like sulfur, nitrogen and many other trace elements — can remove practically all of the pollutants when coal is changed into Synthesis Gas through Coal Gasification. In fact, scientists have ways to remove 99.9% of the sulfur and small dirt particles from the coal gas. Coal Gasification is one of the best ways to clean pollutants out of coal.

Another reason is that the coal gases — carbon monoxide and hydrogen, or simply "Synthesis Gas" — don't have to be burned. They can also be used as valuable chemicals. Scientists have developed chemical reactions that turn carbon monoxide and hydrogen into everything from liquid fuels for cars and trucks to plastic toothbrushes! Today, in Tampa, Florida, and West Terre Haute, Indiana, there are power plants generating electricity through "coal gasification" instead of burning it. At a plant in Kingsport, Tennessee, coal gas is being used to make plastic for photographic film and to make methanol (a fuel that can be burned in automobile engines).

Coal Gasification could be one of the most promising ways to use coal in the future to generate electricity and other valuable products. Yet, it is only one of an entirely new family of energy processes called "Clean Coal Technologies" — technologies that can make fossil fuels future fuels.

Fluidized Bed Combustion

Fluidized beds suspend solid fuels on upward-blowing jets of air during the combustion process. The result is a turbulent mixing of gas and solids. The tumbling action, much like a bubbling fluid, provides more effective chemical reactions and heat transfer. Fluidized bed combustion evolved from efforts to find a combustion process able to control pollutant emissions without external emission controls (such as scrubbers). The technology burns fuel at temperatures of 1,400 to 1,700 degrees F, well below the threshold where Nitrogen Oxides form (at approximately 2,500 degrees F, the nitrogen and oxygen atoms in the combustion air combine to form nitrogen oxide pollutants).

The mixing action of the fluidized bed results brings the flue gases into contact with a sulfur-absorbing chemical, such as limestone or dolomite. More than 95 per cent of the sulfur pollutants in coal can be captured inside the boiler by the sorbent.

Pressurized Fluidized bed combustion (PFBC) builds on earlier work in atmospheric fluidized-bed combustion technology. Atmospheric fluidized bed combustion is crossing over the commercial threshold, with most boiler manufacturers currently offering fluidized bed boilers as a standard package. This success is largely due to the Clean Coal Technology Program and the Energy Department's Fossil Energy and industry partners' R&D.

The popularity of fluidized bed combustion is due largely to the technology's fuel flexibility - almost any combustible material, from coal to municipal waste, can be burned - and the capability of meeting sulfur dioxide and nitrogen oxide emission standards without the need for expensive add-on controls. The Clean Coal Technology Program led to the initial market entry of 1st generation pressurized fluidized bed technology, with an estimated 1000 megawatts of capacity installed worldwide. These systems pressurize the fluidized bed to generate sufficient flue gas energy to drive a gas turbine and operate it in a combined-cycle.

The 1st generation pressurized fluidized bed combustor uses a "bubbling-bed" technology. A relatively stationary fluidized bed is established in the boiler using low air velocities to fluidize the material, and a heat exchanger (boiler tube bundle) immersed in the bed to generate steam. Cyclone separators are used to remove particulate matter from the flue gas prior to entering a gas turbine, which is designed to accept a moderate amount of particulate matter (i.e., "ruggedized").

A 2nd generation pressurized fluidized bed combustor uses "circulating fluidized-bed" technology and a number of efficiency enhancement measures. Circulating fluidized-bed technology has the potential to improve operational characteristics by using higher air flows to entrain and move the bed material, and re-circulating nearly all the bed material with adjacent high-volume, hot cyclone separators. The relatively clean flue gas goes on to the heat exchanger. This approach theoretically simplifies feed design, extends the contact between sorbent and flue gas, reduces likelihood of heat exchanger tube erosion, and improves SO2 capture and combustion efficiency.

A major efficiency enhancing measure for 2nd generation pressurized fluidized bed combustor is the integration of coal gasification to produce Synthesis Gas. This fuel gas is combusted in a topping combustor and adds to the combustor's flue gas energy entering the gas turbine, which is the more efficient portion of the combined cycle. The topping combustor must exhibit flame stability in combusting low-Btu gas and low-NOx emission characteristics. To take maximum advantage of the increasingly efficient commercial gas turbines, the high-energy gas leaving the topping combustor must be nearly free of particulate matter and alkali/sulfur content. Also, releases to the environment from the pressurized fluid bed combustion system must be essentially free of mercury, a soon-to-be regulated hazardous air pollutant.

To reduce cost and carbon dioxide emissions, new sorbents are being evaluated. Sorbent utilization has a major influence on operating costs, and carbon dioxide emissions streams can result in the production and use of alkali-based sorbents.

Efforts are ongoing at the Power Systems Development Facility (PSDF) in Wilsonville, Alabama to ensure critical components and subsystems are ready for demonstration of 2nd generation pressurized fluidized bed combustion. The PSDF is operated by Southern Company Services under DOE contract to conduct cooperative R&D with industry.